GCSE IN A WEEK

SCIENCE

Hannah Kingston
Emma Poole
Caroline Reynolds

Revision Planner

Page	Day	Time (mins)	Title	Exam Board	Date	Time	Completed
4	1	15 mins	How Science Works	AEO			

Biology

Page	Day	Time (mins)	Title	Exam Board	Date	Time	Completed
6	1	15 mins	Healthy Living	A(except Respiration)O			
8	1	15 mins	Central Nervous System	AEO			
10	1	15 mins	The Eye	O			
12	1	15 mins	The Brain	E			
14	1	15 mins	Homeostasis	AEO			
16	1	15 mins	Controlling Fertility	AEO			
18	1	15 mins	Pathogens	AEO			
20	1	15 mins	Natural Defence	AEO			
22	2	15 mins	Artificial Immunity	AEO			
24	2	15 mins	Drugs	AEO			
26	2	15 mins	Genes and Chromosomes	AEO			
28	2	15 mins	Genetic Engineering	AE			
30	2	15 mins	Inheritance and Disease	EO			
32	2	15 mins	Selective Breeding	AE			
34	2	15 mins	The Environment	EO(except Photosynthesis)			
36	2	15 mins	Environmental Damage	AEO			
38	3	15 mins	Ecology and Classification	O			
40	3	15 mins	Adaptation and Competition	AEO			
42	3	15 mins	Evolution	AEO			

Chemistry

Page	Day	Time (mins)	Title	Exam Board	Date	Time	Completed
44	3	15 mins	Air and Air Pollution	AEO			
46	3	15 mins	Crude Oil	AEO			
48	3	15 mins	Food Additives	AEO			
50	3	15 mins	Vegetable Oils	AEO			
52	3	15 mins	Fuels	AEO			
54	4	15 mins	Alkanes and Alkenes	AEO			
56	4	15 mins	Limestone	AEO			
58	4	15 mins	Cosmetics	AEO			
60	4	15 mins	Structure of the Earth	AO			

Page	Day	Time (mins)	Title	Exam Board	Date	Time	Completed

Chemistry cont...

Page	Day	Time (mins)	Title	Exam Board	Date	Time	Completed
62	4	15 mins	New Materials	AEO			
64	4	15 mins	Pollution	AEO			
66	4	15 mins	Useful Metals	AEO			
68	5	15 mins	Iron and Steel	AEO			
70	5	15 mins	Salts	E			
72	5	15 mins	The Periodic Table	AE			
74	5	15 mins	Atomic Structure	AEO			
76	5	15 mins	Chemicals	E			
78	5	15 mins	Rates of Reaction	O			
80	5	15 mins	Energy	AEO			

Physics

Page	Day	Time (mins)	Title	Exam Board	Date	Time	Completed
82	6	15 mins	Producing Current	EO			
84	6	15 mins	Resistance	E			
86	6	15 mins	Power	AEO			
88	6	15 mins	Using Electricity	AEO			
90	6	15 mins	Harnessing Electricity	AEO			
92	6	15 mins	Harnessing Energy 1	AEO			
94	6	15 mins	Harnessing Energy 2	AEO			
96	6	15 mins	Keeping Warm	AEO			
98	6	15 mins	Efficiency	AEO			
100	7	15 mins	Heat	AO			
102	7	15 mins	Heat Calculations	O			
104	7	15 mins	Radiation	AO			
106	7	15 mins	Radiation in our World	AEO			
108	7	15 mins	Waves	AEO			
110	7	15 mins	Waves in our World	AEO			
112	7	15 mins	Using Waves	AEO			
114	7	15 mins	Beyond our Planet	EO			
116	7	15 mins	Communication	AEO			
118	7	15 mins	Space Exploration	O			

How Science Works

Understanding scientific ideas helps us to plan and carry out experiments and become better citizens.

Ideas

Sometimes our opinions are based on our own prejudices; what we personally like or dislike.

At other times, our opinions can be based on scientific evidence. Reliable and valid evidence can be used to back up our own opinions.

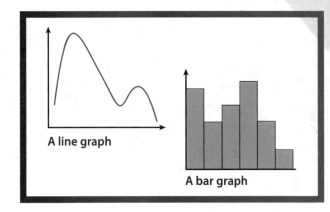

A line graph

A bar graph

Variables

- An **independent** variable is the variable that we choose to change to see what happens.

- A **dependent** variable is the variable that we measure.

- A **continuous** variable, e.g. time or mass, can have any numerical value.

- An **ordered** variable, e.g. small, medium, large, can be listed in order.

- A **discrete** variable can have any value which is a whole number, e.g. 1, 2.

- A **categoric** variable is a variable that can be labelled, e.g. red, blue.

- We use **line graphs** to present data where the independent variable and the dependent variable are both continuous. A line of best fit can be used to show the relationship between variables.

- **Bar graphs** are used to present data when the independent variable is categoric and the dependent variable is continuous.

Evidence

Evidence should be:

- **reliable** (if you do it again you get the same result)

- **accurate** (close to the true value).

Scientists often try to find links between variables.

Links can be:

- causal; a change in one variable produces a change in the other variable

- a chance occurrence

- due to an association, where both of the observed variables are linked by a third variable.

We can use our existing models and ideas to suggest why something happens. This is called a **hypothesis**. We can use this hypothesis to make a **prediction** that can be tested. When the data is collected, if it does not back up our original models and ideas, we need to check that the data is valid. If it is, we need to go back and change our original models and ideas.

Science in Society

Sometimes scientists investigate subjects that have social consequences, e.g. food safety. When this happens, decisions may be based on a combination of the evidence and other factors, such as bias or political considerations.

Although science is helping us to understand more about our world, there are still some questions that we cannot answer, e.g. is there life on other planets? Or, should we clone people?

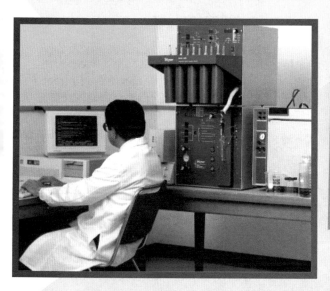

PROGRESS CHECK

1. What is an independent variable?

2. What is a dependent variable?

3. What is an ordered variable?

4. What is a discrete variable?

5. What does 'accurate' mean?

EXAM QUESTION

A student carries out an experiment to find out how the force applied to a spring affects the length of the spring.

a. What is the independent variable?

b. What is the dependent variable?

c. Suggest a variable that must be controlled to make it a fair test.

Healthy Living

To be healthy, you need to eat a balanced diet and take regular exercise. This means avoiding saturated fats and eating only the recommended amount of salt.

Respiration is a chemical process in the cells of your body that provides you with energy. It is important to balance energy intake with the amount of energy used up in exercise, otherwise you could become overweight and suffer from health problems.

proteins, fats, fibre

water, carbohydrates, vitamins and minerals

A balanced diet

A balanced diet consists of:

- **carbohydrates** which provide sugar and starch for energy

- **proteins** which are made up of amino acids and are needed for growth repair and replacement of cells

- **fat** which is made up of fatty acids and glycerol and is needed for making cell membranes and insulation

- **vitamins and minerals** which are essential in small amounts

- **fibre** which keeps food moving through your system and prevents constipation

- **water** which is essential for all the body's chemical reactions.

A total lack of any one important nutrient will cause deficiency diseases and even death. **Kwashiorkor** is a protein deficiency disorder that is common in developing countries where the diet consists of mainly starchy vegetables.

Cholesterol

Cholesterol is made in the liver and is found in the blood. Saturated fats increase blood cholesterol. Monounsaturated fats, however, have little effect on cholesterol and polyunsaturated fats may even help reduce blood cholesterol.

Genetic factors, smoking and alcohol can also contribute to the effects of cholesterol and increase the risk of heart disease.

Statins are drugs that can be used to lower levels of cholesterol in the blood; they are used to treat atherosclerosis and heart disease.

Salt

Salt is needed in small amounts in our diets; an adult needs on average about 6 grams per day but is actually consuming 60% more. Salt contains sodium which is linked to heart disease, high blood pressure and strokes. It is found in high quantities in processed food. Half a gram (0.5 gram) of sodium in foods is considered a large amount, whereas 0.1 gram is a small quantity.

Aerobic Respiration

Respiration is the **breakdown of glucose to make energy using oxygen.**

Energy is needed for all the chemical reactions in the body. During exercise, breathing and pulse rates increase, the arteries supplying the muscles also dilate. This happens in order to deliver oxygen and glucose more quickly to the respiring muscles and to remove carbon dioxide quickly.

The word equation for respiration is:

glucose + oxygen → carbon dioxide + water + **ENERGY**

The symbol equation for respiration is:

$$C_6H_{12}O_6 + 6O_2 \rightarrow 6CO_2 + 6H_2O + \textbf{ENERGY}$$

Anaerobic Respiration and Exercise

Respiration **without oxygen** is called **anaerobic respiration**. It takes place during vigorous exercise, produces much less energy and does not break down glucose completely.

This is the word equation:

glucose → energy + lactic acid

The amount of oxygen needed to remove the lactic acid is called the **oxygen debt**.

PROGRESS CHECK

1. What are the seven food groups that make up a balanced diet?

2. What is the word equation of respiration?

3. In anaerobic respiration, what is produced instead of carbon dioxide and water?

EXAM QUESTION

1. Which type of fat increases blood cholesterol?

2. Describe how cholesterol levels in the blood can be reduced.

3. What other factors can contribute to levels of cholesterol in the blood?

Central Nervous System

The nervous system controls and co-ordinates parts of the body so that they work together at the right time.

The nervous system co-ordinates automatic processes, such as breathing and blinking.

The central nervous system (CNS) consists of the brain and spinal cord connected to different parts of the body by **nerves**.

The body's sense organs are the nose, mouth, ears, skin and eyes. These contain **receptors**. Receptors detect changes in the environment called **stimuli**. The receptors send messages called **nerve impulses** along nerves to the brain and spinal cord in response to stimuli from the environment.

The CNS sends nerve impulses back along nerves to **effectors** which bring about a response. Effectors are muscles that result in movement or they are glands that secrete hormones.

Nerves

Nerves are made up of nerve cells or **neurones**. There are three types of neurone: sensory, motor and relay.

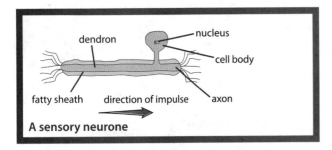

A sensory neurone

The **sensory neurones** receive messages from the receptors and send them to the CNS.

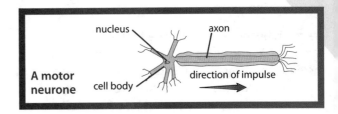

A motor neurone

The **motor neurones** send messages from the CNS to the effectors telling them what to do.

Nerve impulses travel in one direction only. A **relay neurone** connects the sensory neurone to the motor neurone in the CNS.

Synapses

In between the neurones there is a gap called a **synapse**.

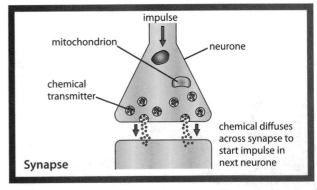

Synapse

When an impulse reaches the end of an axon, a chemical is released.

This chemical diffuses across the gap, starting off an impulse in the next neurone. Synapses can be affected by drugs and alcohol, slowing down synapses or even stopping them.

The Reflex Arc

The reflex response to your CNS and back again can be shown in a diagram called the **reflex arc**.

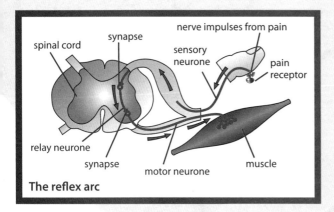

The reflex arc

The reflex arc can be shown in a block diagram:

stimulus → receptor → sensory neurone → relay neurone → motor neurone → effector → response

Reflex and Voluntary Actions

Voluntary actions are things we have to think about – they are under conscious control and they have to be learned, like talking or writing.

Reflex actions produce rapid involuntary responses; they often protect us and other animals from harm. Examples include reflex actions in a newborn baby, the pupils' response to light, the knee jerk reflex and blinking. Simple reflex actions help animals survive as they respond to a stimulus such as smelling and finding food or avoidance of predators.

PROGRESS CHECK

1. What does the abbreviation CNS stand for?

2. Name the **three** types of neurone.

3. What is the difference between a voluntary and reflex action?

EXAM QUESTION

1. Fill in the following gaps to show the path taken by a nerve impulse:

   Stimulus → _____ → Sensory neurone → _____ neurone → Motor neurone → _____ → Response

2. How is the nerve impulse transmitted from neurone to neurone?

3. What do we call a change in the environment that would generate a nerve impulse?

The Eye

The eye is one of the human sense organs. Parts of the eye control the amount of light entering it and other parts control our focus on near and distant objects.

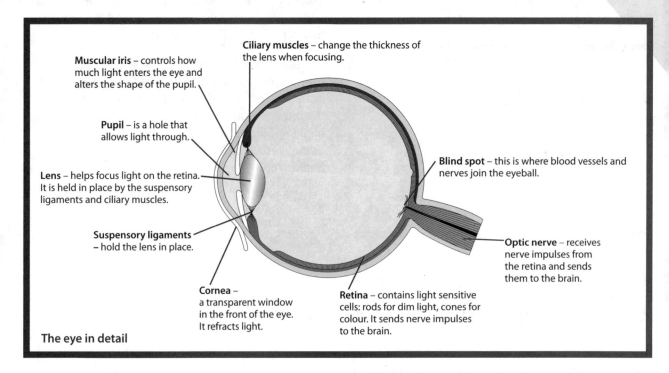

Ciliary muscles – change the thickness of the lens when focusing.

Muscular iris – controls how much light enters the eye and alters the shape of the pupil.

Pupil – is a hole that allows light through.

Lens – helps focus light on the retina. It is held in place by the suspensory ligaments and ciliary muscles.

Suspensory ligaments – hold the lens in place.

Cornea – a transparent window in the front of the eye. It refracts light.

Blind spot – this is where blood vessels and nerves join the eyeball.

Optic nerve – receives nerve impulses from the retina and sends them to the brain.

Retina – contains light sensitive cells: rods for dim light, cones for colour. It sends nerve impulses to the brain.

The eye in detail

Adjusting to Light and Dark

Bright light

- ■ Circular muscles contract. Radial muscle relaxes.

- ■ The iris closes and makes the **pupil smaller**.

Dim light

- ■ Radial muscles contract. Circular muscles relax.

- ■ The iris opens and makes the **pupil bigger**.

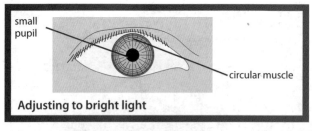

small pupil

circular muscle

Adjusting to bright light

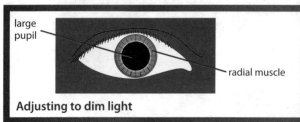

large pupil

radial muscle

Adjusting to dim light

Focusing on Objects

Light passes through the cornea and lens and the object is focused on the retina. This is called **accommodation**.

	Ciliary muscles	Suspensory ligaments	Lens shape
Near objects	Contract	Slacken	Fat and round
Far objects	Relax	Contract	Thin and flat

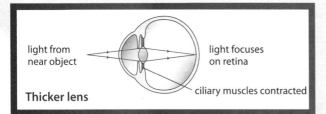

light from near object — light focuses on retina

Thicker lens — ciliary muscles contracted

Vision

Humans and many hunting animals have **binocular vision**. This means that our eyes are facing forward which enables us to judge distances and depth effectively. Cows, horses and other prey animals have **monocular vision** with eyes on the side of their heads. This allows them to have a wider field of view.

Problems with Vision

Short-sightedness results when the eyeball is too long. This means that light is focused too far in front of the retina. Sufferers can see near objects but not distant ones.

Long-sightedness is when the eyeball is too short and distant objects can be seen, but not close-up ones.

Treatment for both short- and long-sightedness involves contact lenses and glasses with different-shaped lenses, or corneal surgery.

Red–green colour blindness is an inherited condition which affects more males than females. It is caused by the cones not functioning correctly.

Older people sometimes suffer from poor accommodation; they cannot focus quickly enough from close to distant objects because of weak ciliary muscles or stiff suspensory ligaments. This can cause problems judging distances when driving.

PROGRESS CHECK

1. What is the difference between binocular vision and monocular vision?

2. Name the muscles that control the size of the lens.

3. What shape is the lens when focusing on near objects?

? EXAM QUESTION

1. Match each of the following parts of the eye with its function.

   Iris   Changes shape to focus
   Lens   Holds the lens in place
   Suspensory ligaments   Controls how much light enters the eye

2. Describe how the eye accommodates.

3. Jean is an elderly lady who has difficulty with accommodation. What causes this difficulty and why might it be a problem?

The Brain

The brain is situated at the top of the spinal cord and is protected by the skull. The brain, spinal cord and neurones make up the **central nervous system**.

The **cerebral cortex** makes up the outer layer of the brain and is in two halves called the cerebral hemispheres. These are made up of lobes.

The **medulla** is the part of the brain that attaches to the spinal cord. It controls automatic actions such as breathing and heart rate. The **cerebellum** controls our co-ordination and balance.

The Brain and Learning

The brain works by sending electrical impulses received from the sense organs to the muscles. It co-ordinates the response.

In mammals, the brain involves billions of neurones that allow learning by experience and behaviour through interaction with the environment. This results in nerve pathways forming in the brain. When we learn from experience, some pathways in the brain become more likely to transmit impulses than others, so it is easier to learn through repetition.

The way in which humans learn language has been long debated by linguists and child psychologists. Some say that there is a crucial period of language acquisition that ends when a child is around 12 years of age.

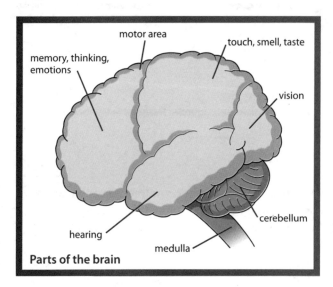

Parts of the brain

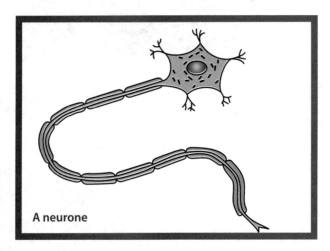

A neurone

Brain Disorders

Strokes

Strokes occur when the blood supply to the brain is stopped. The neurones then start to die and this leads to paralysis and loss of speech. The causes are blockage of blood vessels in the brain.

Symptoms include weakness/numbness in the face, leg, arm, one side of the body, and loss of vision, difficulty in speaking, headache and dizziness.

Strokes are linked to high blood pressure, smoking, heart disease and diabetes.

Epilepsy

Epilepsy is the disruption of electrical activity in the brain that causes abnormal functioning and seizures. Seizures prevent the brain from interpreting and processing signals such as sight, hearing and muscle control.

There are two types of seizures: **grand mal seizures** where sufferers have convulsions, and **petit mal seizures** which are non-convulsive.

Causes of epilepsy include head injuries, strokes, brain tumours or infections such as meningitis. The attacks can be brought on by stress, lack of sleep, flashing lights or sounds and low blood sugar levels.

Parkinson's Disease

Parkinson's disease is a chronic progressive movement disorder in which the brain degenerates. The cause is unknown. Symptoms include tremors, rigidity, slow movement, poor balance and difficulty walking.

Brain Tumours

Brain tumours are caused by uncontrollable growth of cells. They can be malignant and cancerous, or benign but still cause pressure on the skull. The causes are unknown, but the risks increase through exposure to radiation or chemicals, or if the immune system is weakened through illness.

PROGRESS CHECK

1. Which part of the brain controls breathing and heart rate?

2. What disorder disrupts the electrical activity of the brain?

3. Which part of the brain controls co-ordination and balance?

EXAM QUESTION

1. Describe how strokes affect the normal functioning of the brain.

2. Name **two** other disorders of the brain.

3. We learn through experience and behaviour. How does repeating an experience help us learn?

Homeostasis

The nervous system and hormones enable us to respond to external changes in the environment by monitoring and changing our internal environment.

Internal Environment

Hormones are chemical messengers produced by glands known as endocrine glands. They travel in the blood.

Hormonal effects tend to be slow, long lasting and affect a number of organs. Nervous control is much quicker, does not last very long, and is confined to one area.

Our internal environment changes very little and stays at a safe level. This is called **homeostasis**.

Internal conditions of the body which are controlled include:

- water content of the blood
- ion content of the blood
- temperature
- blood sugar levels.

The Pancreas and Homeostasis

The pancreas **maintains the level of glucose (sugar) in the blood** so that there is enough for respiration.

Diabetes results when the **pancreas fails to make enough of the hormone insulin**. This can make the sufferer tired and thirsty. If untreated, it can lead to weight loss and even death.

Diabetes can be controlled by **attention to diet** – a low glucose diet can be all that is needed. In more severe cases, diabetics have to **inject themselves with insulin** before meals.

The pancreas secretes two hormones: **insulin and glucagon**. The liver responds to insulin, takes up excess glucose and stores it as **glycogen**.

Glucagon stimulates the conversion of stored glycogen in the liver back into glucose. This is an example of **negative feedback**.

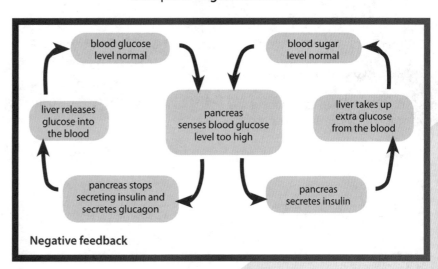

Negative feedback

Body Temperature

Warm-blooded animals have mechanisms that can keep body temperature constant. The hypothalamus in the brain controls this. Inside the body, the temperature stays around the same, 37°C.

When it is hot:

- blood vessels at the surface of the skin widen allowing more blood to flow to the surface. This is called **vasodilation**

- heat radiates from the skin and is lost

- sweat glands secrete sweat

- sweat evaporates from the skin and takes away heat energy.

When it is cold:

- many warm-blooded animals have a thick layer of fat beneath their skin for insulation

- blood vessels at the surface of the skin contract so that very little blood reaches to the surface. This is called **vasoconstriction**

- very little heat is lost by radiation

- muscles contract quickly (shivering) which produces extra heat

- sweat glands stop producing sweat

- increased respiration helps generate heat, as does exercise.

Extremes of Temperature

If the outside temperature is extremely cold and the body temperature falls dramatically, the control centre in the brain stops working. If it drops too low, sufferers may slip into a coma and die if no action is taken. This is called hypothermia. Babies and old people are more susceptible to hypothermia.

Alternatively, if it becomes too hot, then it can lead to heat stroke.

PROGRESS CHECK

1. What is homeostasis?

2. Which hormone (i) raises (ii) lowers blood sugar level?

3. What causes diabetes?

EXAM QUESTION

Tom was walking outside in the cold; he looked pale and was shivering.

a. Explain how his body was working to try to keep him warm.

b. Which part of the brain monitors body temperature?

c. Circle the correct answer.

   Our body temperature is maintained at around:

   37°C   40°C   35°C

Controlling Fertility

The menstrual cycle lasts approximately **28 days**. It consists of a **menstrual bleed and ovulation** – the release of an egg.

Menstrual Cycle Stages

Hormones control the whole cycle. Ovaries secrete the hormones **progesterone and oestrogen**.

- **Days 1–5**, a menstrual bleed (a period) occurs. The lining of the uterus breaks down. This is caused by lack of progesterone.

- **Days 5–14**, oestrogen is released and the uterus lining builds up again. Oestrogen stimulates egg development and release of the egg from the ovaries – this is called ovulation.

- **Days 14–28**, progesterone is released which maintains the uterus lining. If no fertilisation occurs, progesterone production stops.

- **Days 28–5**, the cycle begins again.

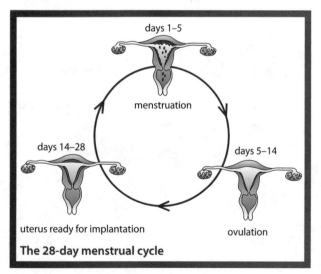

The 28-day menstrual cycle

The Pituitary Gland

The hormones released from the ovaries are controlled by the **pituitary gland** which is situated at the base of the brain.

The pituitary gland secretes two hormones, follicle stimulating hormone and luteinising hormone, written as **FSH** and **LH**. The hormones interact to control the menstrual cycle.

Progesterone is released which maintains the uterus lining. Oestrogen also keeps the uterus lining thick ready for pregnancy. If no pregnancy occurs, progesterone production stops and the cycle begins again.

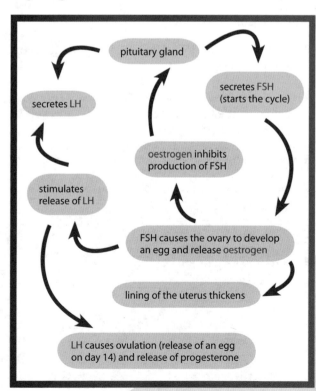

Controlling Fertility

Fertility in women can be controlled using artificial hormones. FSH can be administered as a **fertility drug** to women whose own production is too low to stimulate eggs to mature. Sometimes the use of FSH can result in multiple births.

Oestrogen can be used as an **oral contraceptive** to inhibit FSH so that no eggs mature.

IVF

In vitro fertilisation (IVF) is a treatment for infertile couples. It involves extracting the eggs and sperm and fertilising them outside the body. The cells that develop are then implanted in the womb for growth and development into an embryo.

It is an expensive process and is not always successful. **FSH** is the hormone used to stimulate egg production.

There are some social and ethical implications of IVF, particularly its use with older couples. Some clinics set an age limit; they feel the chances of success fall with increasing age, and also the possibility of birth defects increases with older eggs.

IVF can create many fertilised embryos; this raises the question of what to do with the ones that are not implanted back in the womb. Some couples look at freezing them for later use, but what about the rest? Should they be thrown away? Used for medical research? What if the couples split up? Who owns the embryos?

PROGRESS CHECK

1. Where are the hormones oestrogen and progesterone made?

2. Which **two** hormones are produced by the pituitary gland?

3. What is ovulation?

EXAM QUESTION

1. Fill in the gaps using the following words:

   oestrogen　　**progesterone**　　**FSH**　　**LH**

   _____ begins the menstrual cycle and causes the ovary to mature an egg and release _____. Oestrogen inhibits FSH but causes the release of _____. LH causes ovulation and the release of _____ which maintains the uterus lining.

2. Which hormone can be used as an oral contraceptive?

3. Which hormone can be used in fertility treatment?

Pathogens

Micro-organisms include bacteria, viruses, fungi and single-celled organisms called protozoa. If they get inside you and make you feel ill, they are called **pathogens**.

Diseases are spread through:

- **contact** with infected people or objects used by infected people, e.g. athlete's foot

- the **air**, e.g. flu, colds and pneumonia

- infected **food and drink**, e.g. cholera from infected drinking water and salmonella food poisoning.

Bacteria

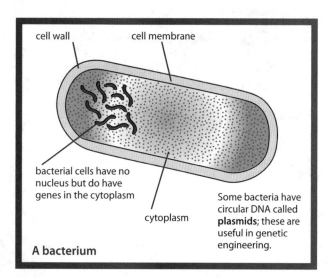

cell wall

cell membrane

bacterial cells have no nucleus but do have genes in the cytoplasm

cytoplasm

Some bacteria have circular DNA called **plasmids**; these are useful in genetic engineering.

A bacterium

Bacteria are living organisms that feed, move and carry out respiration. They reproduce rapidly and produce exact copies of themselves. They are good at surviving in unfavourable conditions.

How Bacteria Cause Disease

Bacteria **destroy living tissue**. For example, tuberculosis destroys lung tissue.

Bacteria can produce poisons called **toxins**. Food poisoning is an example.

Viruses

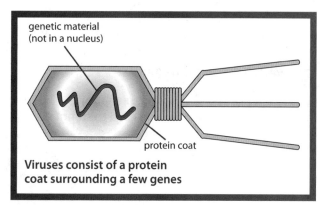

genetic material (not in a nucleus)

protein coat

Viruses consist of a protein coat surrounding a few genes

Viruses are much smaller than bacteria. They do not feed, move, respire or grow: they just reproduce. Viruses can only survive inside the cells of a living organism. They reproduce inside the cells and release thousands of new viruses to infect new cells, killing the cell in the process. Examples of viruses which cause disease are HIV, flu, chicken pox and measles.

Fungi

Fungi cause athlete's foot and ringworm. They reproduce by **making spores** that can be carried from person to person.

Most fungi are useful as decomposers. Yeast is used in bread, beer and wine making.

Protozoa

Protozoa are tiny and made up of only one cell; some types may cause dysentery.

How do pathogens get in?

Pathogens have to enter our body before they can do any harm. The pathogens can enter in the following ways:

- skin – when **damaged**
- digestive system – via food and drink
- respiratory system – viruses are breathed in
- reproductive system – through sexual intercourse.

Vectors

Some pathogens rely on **vectors** to transfer them from one organism to another, e.g. **mosquitoes**.

A mosquito that is carrying the malarial parasite infects a person by injecting the *Plasmodium* parasite into the person's bloodstream when it bites. A parasite is an organism that lives off another without any benefit to the host.

Cancer

Cancer is a condition caused when body cells grow out of control and become a mass of cells known as a tumour. Tumours that stop growing are **benign**; ones that invade the surrounding tissues and organs are called **malignant**.

The most common cancer in men is prostate cancer. There is some evidence that eating a low fat diet and taking the supplement selenium reduces the risk.

Breast cancer may affect up to one in nine women in their life and 1% of males. It often runs in families.

Skin cancer is caused by the over exposure to ultraviolet (UV) sunlight or sun beds.

PROGRESS CHECK

1. Name the **four** types of micro-organism.

2. What do we call micro-organisms that cause disease?

3. What is a vector? Give an example.

EXAM QUESTION

1. What type of cell is this?

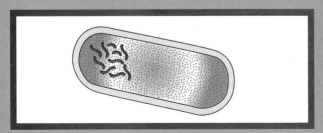

2. Explain how it causes disease.

3. Name **two** other disease-causing micro-organisms.

Natural Defence

The human body has many methods of **preventing pathogens** from entering the body.

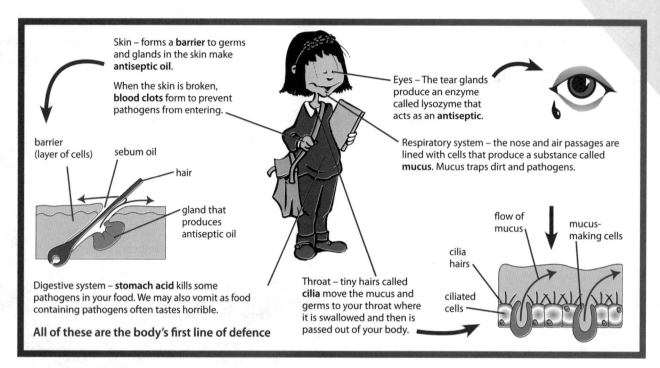

Skin – forms a **barrier** to germs and glands in the skin make **antiseptic oil**.

When the skin is broken, **blood clots** form to prevent pathogens from entering.

barrier (layer of cells)

sebum oil

hair

gland that produces antiseptic oil

Eyes – The tear glands produce an enzyme called lysozyme that acts as an **antiseptic**.

Respiratory system – the nose and air passages are lined with cells that produce a substance called **mucus**. Mucus traps dirt and pathogens.

flow of mucus

mucus-making cells

cilia hairs

ciliated cells

Digestive system – **stomach acid** kills some pathogens in your food. We may also vomit as food containing pathogens often tastes horrible.

Throat – tiny hairs called **cilia** move the mucus and germs to your throat where it is swallowed and then is passed out of your body.

All of these are the body's first line of defence

Preventing the Spread of Germs

We can prevent the spread of germs by:

- **sterilising** equipment used in food preparation or operating theatres by heating them to 120°C

- **disinfecting** work surfaces and areas like toilets

- using **antiseptics** which can kill pathogens if we cut ourselves

- having **general good hygiene** as this is important in preventing the spread of disease.

In 1847, Dr. Semmelweiss suggested that washing hands prevents infection. He noticed that there seemed to be a spread of childhood fever that was significantly reduced when hands were washed between patients. He suggested the modern-day idea of disinfecting hands and instruments.

The Immune System Response

If the pathogens do get into the body, **white blood cells** travelling in the blood are activated.

- White blood cells make chemicals called **antitoxins** that destroy the **toxins** produced by bacteria and prevent inflammation.

- White blood cells called **phagocytes** engulf the odd bacteria or viruses before they have a chance to do any harm.

This is the body's **second line of defence** and is non-specific.

However, if large numbers of toxins enter, then a type of white blood cell called **lymphocytes** become involved.

This is the **third line of defence** and is called the **specific immune response**.

All pathogens have chemicals on their surface called **antigens**; these are recognised by lymphocytes. They produce chemical **antibodies** that attach to antigens and destroy them in various ways, like clumping them together so that phagocytes can engulf and destroy them. Antibodies are made **specifically** to fight a particular antigen.

Natural Immunity

Making antibodies takes time. Initially, an infected person feels ill, and then gets better because the disease is destroyed by the white blood cells. Once a particular antibody is made, a few stay in the body and act as a memory. If the same disease enters the body the antibodies are much quicker at multiplying and destroying it and the person may feel no symptoms. The person is then **immune** to that disease.

PROGRESS CHECK

1. What chemicals do white blood cells produce to fight toxins?

2. How does the body recognise foreign bacteria and viruses?

3. Name **four** ways of preventing the spread of pathogens.

EXAM QUESTION

1. Rachel has cut her finger and it has become infected with bacteria. Explain how phagocytes can help get rid of the infection.

2. Explain briefly how lymphocytes fight infections.

3. Why is this called the specific immune response?

Artificial Immunity

There are various ways of treating disease and infection once they get past the body's natural defences.

Vaccines

A vaccine contains dead or harmless pathogens that still have antigens. White blood cells respond as if they were alive by multiplying and producing antibodies.

A vaccine acts like an advanced warning, so that if the person is infected by the pathogen again, the white blood cells can respond immediately.

Vaccinations are **passive immunity** as the body produces its own antibodies. An injection of ready-made antibodies is called **active immunity**.

Vaccines help prevent the spread of disease and epidemics, but people have the right to choose whether to be vaccinated or not.

The vaccine to treat MMR (measles, mumps and rubella) is a viral vaccine that has caused much controversy. This is because of the concern about the side effects of using a triple vaccine instead of three separate ones. There is also the problem of possible long-term side effects.

New vaccines against 'flu are needed regularly as the virus constantly changes.

Antibiotics

Sometimes bacteria breach the body's defences and reproduce successfully. When this happens, outside help is needed in the form of **antibiotics**.

Penicillin was the first form of antibiotic. It is made from a mould called *Penicillium notatum*. Antibiotics can only treat infections caused by bacteria; they cannot treat infections caused by viruses. The body has to fight a virus on its own.

Mutations

Antibiotics kill most bacteria, but as we continue to use them, bacteria are becoming **resistant** to them. New antibiotics are constantly needed.

MRSA is a 'superbug' that thrives in unhygienic conditions, is highly contagious and can mutate. Places such as hospitals need scrupulous hygiene to try and prevent it developing and spreading and they need to avoid the overuse of antibiotics.

Drug Testing

New drugs and medical treatments have to be extensively tested and trialled before being used. They are tested in the laboratory before being tested on human volunteers. They are first tested on healthy volunteers to test for safety, and then people with the illness to test for effectiveness.

The tests are normally **blind** or **double blind**, which means neither the patients nor the doctors know which are being treated. The patients still receive their normal treatment for their disease. Placebos (a control group who receive a dummy drug) are not often used as patients must still receive treatment.

Sometimes the tests fail. This happened with a drug called thalidomide which was developed as a sleeping pill. It was found to be effective at treating morning sickness; however it had not been tested for this use.

This drug caused severe limb abnormalities in babies born to mothers taking the drug. It was subsequently banned. It is now being used to treat leprosy and some forms of cancer.

PROGRESS CHECK

1. What was the first form of antibiotic called?

2. How do bacteria and viruses become resistant to antibiotics?

3. What does the term 'double blind' mean?

EXAM QUESTION

1. Laura has the cold virus, but the doctor won't supply her with antibiotics. Explain why.

2. Laura has had the flu vaccination. Explain why this doesn't prevent her from getting a cold.

3. Is a vaccine an example of passive or active immunity?

Drugs

Drugs are powerful chemicals that alter the way the body works.

Why are Drugs Dangerous?

Drugs can become addictive; people become dependent on them and suffer from withdrawal symptoms when they stop taking them.

Some drugs taken for recreational use are illegal, but other drugs, such as alcohol and smoking, are legal. The overall impact on health from legal drugs can be greater, as more people take them than the illegal drugs.

Drugs fall into four main groups:

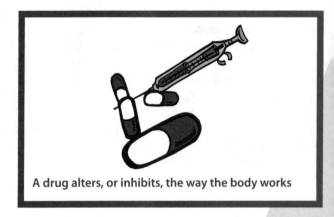

A drug alters, or inhibits, the way the body works

1. Sedatives

These drugs slow down the brain. Tranquillisers and sleeping pills are examples of sedatives. Sedatives seriously alter reaction times and affect a person's judgement of speed and distances when driving.

2. Painkillers

These drugs suppress the pain sensors in the brain.

Paracetamol, aspirin, heroin and morphine are examples of painkillers. There are approximately 130 deaths each year in England and Wales from paracetamol overdose.

3. Hallucinogens

These drugs can cause hallucinations – people see or hear things that do not exist. Hallucinogens can lead to fatal accidents.

Examples of hallucinogens are ecstasy, LSD and cannabis.

4. Stimulants

These drugs speed up the brain and nervous system. They make you more alert and awake.

Examples include amphetamines, cocaine, and the less harmful caffeine in tea and coffee.

Drugs and the Law

Drugs are classified in law as class A, B or C. Class A drugs, such as heroin, carry the most severe penalties if a person is caught in possession of them. Supply of the drug can lead to life imprisonment.

Class B drugs, such as amphetamines, still carry severe penalties. Recently cannabis has become a class C drug carrying less harsh penalties if caught in possession. However, if cannabis is in possession with intent to supply, the penalty is up to 14 years' imprisonment.

The cannabis debate still continues about whether it is harmful, addictive or leads on to the use of harder drugs such as heroin. At present health professionals cannot agree.

Solvents

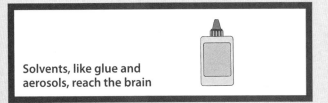

Solvents, like glue and aerosols, reach the brain

Solvents include everyday products like glue and aerosols. Solvent fumes are inhaled and absorbed by the lungs. They soon reach the brain and slow down breathing and heart rates. Solvents also damage the kidneys and liver.

Alcohol

Alcohol is a **depressant** and reduces the activity of the brain and nervous system. It is absorbed through the gut and taken to the brain in the blood.

Alcohol damages neurones in the brain that causes irreversible brain damage.

The liver breaks down alcohol, but in excess, it has a very damaging effect on the liver causing a disease called **cirrhosis**.

Each of these drinks contains one unit of alcohol

1 glass of sherry

$\frac{1}{2}$ pint cider (0.3 litre)

1 glass of wine

$\frac{1}{2}$ pint beer (0.3 litre)

1 single whisky

Smoking

Smoking: a proven cause of health problems

Tobacco contains many harmful chemicals; **nicotine** is an addictive substance and a mild stimulant, **tar** is known to contain carcinogens that contribute to cancer and **carbon monoxide** prevents the red blood cells from carrying oxygen.

If pregnant women smoke, then carbon monoxide deprives the foetus of oxygen and can lead to a low birth mass. Some diseases caused by smoking include **emphysema, bronchitis, heart and blood vessel problems,** and **lung cancer**.

PROGRESS CHECK

1. What parts of the body are affected by solvents?

2. Name **three** chemicals contained in tobacco.

3. What diseases may be caused by smoking?

? EXAM QUESTION

For each of the following drugs explain one effect on the body:

a. solvents

b. alcohol

c. stimulants

d. sedatives

Genes and Chromosomes

Inside nearly all cells is a **nucleus** containing instructions that control all our inherited characteristics.

The instructions are carried on **chromosomes** that occur in pairs – one each from the mother and father. Genes on chromosomes control particular **characteristics**.

Inside human cells, there are **23 pairs** or **46 chromosomes**. The cell is called a **diploid cell**. Each chromosome is made up of a long-stranded molecule called **DNA**. A gene is a section of DNA.

Proteins and enzymes control all our characteristics. Genes are chemical instructions that code for a particular protein or enzyme and therefore our characteristics. There is a **pair of genes** for each feature.

The name we give to different versions of a gene is **alleles**.

DNA molecule

A DNA molecule makes up the arms of a chromosome. The molecule is coiled together to form a **double helix shape** and is joined by chemical bases, like rungs in a ladder. DNA has the ability to copy itself exactly so that any new cells made have exactly the same genetic information. Each gene contains a different sequence of bases.

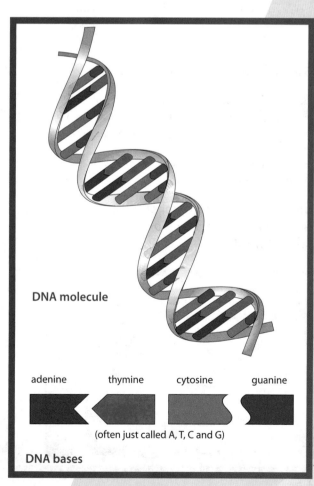

DNA molecule

adenine   thymine   cytosine   guanine

(often just called A, T, C and G)

DNA bases

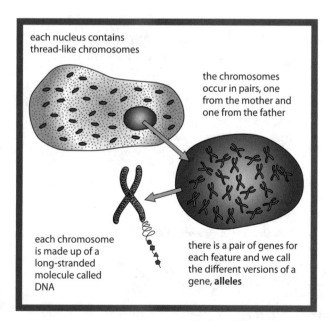

each nucleus contains thread-like chromosomes

the chromosomes occur in pairs, one from the mother and one from the father

each chromosome is made up of a long-stranded molecule called DNA

there is a pair of genes for each feature and we call the different versions of a gene, alleles

Inheritance

In order to inherit characteristics from the parent DNA, a form of reproduction needs to take place:

- **Asexual reproduction** involves only one parent; the offspring have exact copies of the parental genes. (They are clones.) There is no fusing (joining) of the parental gametes (sperm and eggs).

- **Sexual reproduction** involves fertilisation and two parents. The gametes' nuclei fuse and the genes are passed on to the offspring. The offspring are not genetically identical.

Each gamete contains half the number of chromosomes (23). Upon fertilisation, the number becomes 46, a full set of 23 pairs, to produce a unique individual.

Variation between individuals is partly due to the inheritance of genes from the parents but also the environment in which the offspring live and grow.

Some human and animal characteristics are controlled by genes and are inherited, for example, nose shape, blood group, natural hair colour, eye colour and gender. Other characteristics are controlled by the environment, for example scars or language.

Mutations

A mutation is a change in the chemical structure of a gene or chromosome which alters the way an organism develops. It may happen for no reason, or there might be a cause.

Mutations occur naturally in the environment, for example, new strains of the 'flu virus are always appearing. Mutations that occur to body cells are not inherited; they are only harmful to the person whose body cells are altered.

Mutations that occur in reproductive cells are inherited; the child will develop abnormally or die at an early age.

Some mutations that are inherited are beneficial and form the basis of evolution.

PROGRESS CHECK

1. How many chromosomes does a human body cell have?

2. In what circumstances are mutations inherited?

3. Which type of reproduction formed you? What else has an effect on how you look and behave?

EXAM QUESTION

Match the following words to their best definition.

DNA	Alternative form of a gene
Alleles	Chromosomes are made up of this
Genes	There are 23 of these in human gametes
Chromosomes	Instructions for characteristics

Genetic Engineering

Genetic engineering is the process in which genes from one organism are removed and inserted into the cells of another.

Making Insulin

Many diseases are caused when the body cannot make a particular protein.

Genetic engineering has been used to treat people with diabetes by the production of the protein called insulin. This involves the use of plasmids (circular DNA) from bacteria.

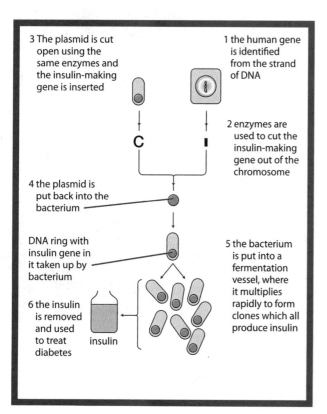

3 The plasmid is cut open using the same enzymes and the insulin-making gene is inserted

1 the human gene is identified from the strand of DNA

C

2 enzymes are used to cut the insulin-making gene out of the chromosome

4 the plasmid is put back into the bacterium

DNA ring with insulin gene in it taken up by bacterium

5 the bacterium is put into a fermentation vessel, where it multiplies rapidly to form clones which all produce insulin

6 the insulin is removed and used to treat diabetes

insulin

The Benefits of Genetic Engineering

Genetic engineering benefits industry, medicine and agriculture. It benefits agriculture by developing plants that are resistant to pests and diseases, and can grow in less desirable conditions. Wheat and other crops have been developed to utilise nitrogen from the air directly in order to produce proteins without the need for fertilisers.

Fruits are now able to stay fresh for longer. Animals are engineered to produce chemicals in their milk such as drugs and human antibodies.

Risks of Genetic Engineering

Manipulating bacteria for use in producing proteins could result in previously harmless bacteria mutating into a disease causing bacteria.

There is concern that GM crops damage human health by causing allergies and there is the notion of eating 'foreign' DNA. There is also concern for the nutritional quality of GM crops. It is argued that GM crops damage the environment, and there is the risk of crops transferring their resistance to herbicide gene to weeds.

Genetic engineering is seen by many as manipulating the 'stuff of life'. Is it morally and ethically wrong to interfere with nature?

Tomatoes can be genetically engineered to stay fresh longer by inserting a gene from fish into the cells.

GM Crops

Presently there are no genetically modified crops grown in the UK, though they have been grown for research and development purposes. There are plans to grow them in the UK in 2008, however.

Countries that do grow GM crops include the USA, Argentina, Canada and China. The quantities and varieties of crops that they grow vary, but crops include corn, cotton and soya bean.

Gene Therapy

It may be possible to treat inherited diseases such as **cystic fibrosis**. People with the disease could be cured if the correct gene is inserted into their body cells. The cells that need the gene, however, are in many parts of the body.

Gene therapy is potentially a way forward in curing fatal diseases, but it poses risks as inserting genes into human cells may make them cancerous.

Human Genome Project

This project was finalised in 2003. It took scientists from all around the world 10 years to complete. It aimed to identify all the genes in human DNA and study them.

The benefits of the Human Genome Project include improved diagnosis for disease and earlier detection of genetic diseases in families such as breast cancer and Alzheimer's disease. The use of DNA in forensic science has also been improved.

PROGRESS CHECK

1. Give an outline of the Human Genome Project.

2. What is genetic engineering?

3. What is gene therapy?

EXAM QUESTION

1. Where is the human gene for insulin found?

2. Explain briefly the stages involved in producing insulin from bacteria.

3. What is the potential problem with using bacteria for genetic engineering?

Inheritance and Disease

Genes pass on characteristics from one generation to the next. Sometimes 'faulty' genes are inherited that cause genetic diseases.

Cystic Fibrosis

Cystic fibrosis is the commonest inherited disease in Britain. About 1 in 2000 children born in Britain each year has cystic fibrosis.

Cystic fibrosis is caused by a **recessive allele (c)**, carried by about one person in 20. People who have the genotype (Cc) are said to be **carriers**, with no ill effects. Only people who have the genotype (cc) are affected.

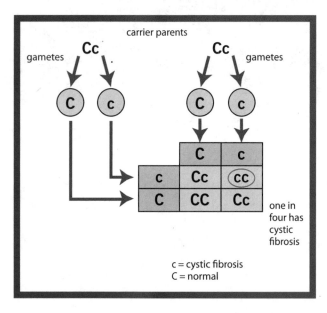

carrier parents

gametes Cc   Cc gametes

C   c   C   c

	C	c
c	Cc	cc
C	CC	Cc

one in four has cystic fibrosis

c = cystic fibrosis
C = normal

Cystic fibrosis sufferers produce large amounts of thick, sticky mucus that can block air passages and digestive tubes. This causes difficulty in breathing and absorbing food. The mucus also encourages bacteria to grow, which causes chest infections. There is still no cure; treatment involves **physiotherapy** to remove some of the mucus, and **strong antibiotics** to treat infections.

Sickle Cell Anaemia

Sickle cell anaemia is caused by a **recessive allele(s)**. Two carrier parents have a one in four chance of having a child with the disease and the genotype(**ss**).

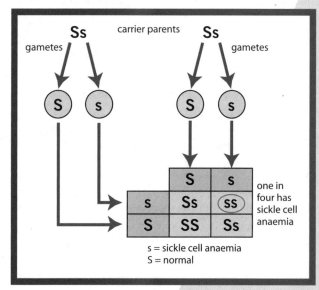

Ss   carrier parents   Ss
gametes   gametes

S   s   S   s

	S	s
s	Ss	ss
S	SS	Ss

one in four has sickle cell anaemia

s = sickle cell anaemia
S = normal

Sickle cell is a disease of the **blood**. The red blood cells are an abnormal shape. The sickle shape affects the oxygen carrying capacity of the blood and becomes stuck in the capillaries.

It is an extremely painful disease and sufferers usually die at an early age. There is no cure and, even though sufferers may die before they can reproduce, the disease has not disappeared, especially in malarial regions such as Africa. This is because **carriers** with the genotype **(Ss)** are immune to malaria.

Genetics

Genetics is the study of how information is passed on through generations.

Definitions

- **Gene**: the unit of inheritance carried on chromosomes. Alternative forms of a gene are called **alleles**.

- **Recessive**: only has an effect in the homozygous recessive condition.

- **Dominant** means that that allele is stronger and has an effect in the heterozygous condition.

- **Genotype** is the type of alleles carried by an organism.

- **Phenotype** is what the organism physically looks like, the result of the genotype.

- If an organism has both alleles the same they are **homozygous dominant** or **homozygous recessive**.

- If an organism has different alleles they are **heterozygous**.

A Worked Example

Letters are used to represent alleles, upper case for dominant characteristics and lower case for recessive characteristics.

The allele for brown eyes is **dominant** (B) to the **recessive** blue eyes allele (b).

If the mother and father are **heterozygous** for eye colour, then they have the genotype **Bb**. What colour eyes will their children have?

We can show the possible outcomes using a **genetic diagram:**

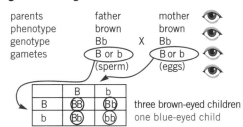

parents		father		mother	
phenotype		brown		brown	
genotype		Bb	X	Bb	
gametes		B or b (sperm)		B or b (eggs)	

	B	b
B	BB	Bb
b	Bb	bb

three brown-eyed children
one blue-eyed child

This gives a 3:1 ratio of brown to blue eyes.

PROGRESS CHECK

1. How is cystic fibrosis inherited?

2. What is sickle cell anaemia?

3. When can being a carrier of sickle cell be an advantage?

? EXAM QUESTION

Tongue rolling is dominant to non-tongue rolling. David has the genotype Rr and Helen has the genotype Rr.

a. What are the possible genotypes of their children? Use a genetic diagram in your answer.

b. What are the possible genotypes for tongue rolling and non-tongue rolling?

Selective Breeding

Selective breeding is where humans try to improve animals and plants by breeding from the best individuals and hoping for improved stock. It can also be called **artificial selection**.

Selective Breeding in Animals

Dogs have been selectively bred over many years to produce the variety of breeds that we have today. Cows have been selectively bred to produce a greater quantity of milk. Beef cattle have been bred to produce better meat.

New techniques have been developed to produce more offspring in a shorter space of time.

Embryo Transplants

The process is as follows:

1. Sperm is taken from the best bull.

2. The best cow is given hormones to produce lots of eggs.

3. The eggs are removed from the cow and fertilised in a petri dish.

4. The embryos are allowed to develop but are then split apart to form clones before they become specialised.

5. The embryos are implanted into other cows, called surrogates, where they grow into the desired offspring.

Advantages are: the sperm and the eggs can be frozen for use at a later date; a large number of offspring can be produced from one bull and one cow.

Selective Breeding in Plants

Selectively bred individuals may not always produce the desired characteristics if the breeding involves sexual reproduction.

With plants, this can be overcome by producing clones using asexual reproduction. Many plants reproduce asexually on their own, such as strawberry plants that produce runners. Other examples are potatoes, onions and *Chlorophytum* (spider plant). Plants also produce sexually, attracting insects for pollination.

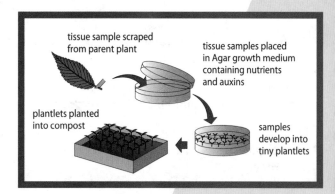

tissue sample scraped from parent plant

tissue samples placed in Agar growth medium containing nutrients and auxins

plantlets planted into compost

samples develop into tiny plantlets

Gardeners can produce identical plants by taking **cuttings** from an original plant. The cuttings are dipped in hormone powder and then grown into new plants.

Tissue culture is a technique used by commercial plant breeders. A few plant cells are taken and a new plant is grown from them, using a special growth medium containing hormones. The advantages are that new plants can be grown quickly and cheaply all year round with special properties such as resistance to diseases.

Problems

If animals or plants are continually bred from the same best animals or plants, then the animals and plants will all be very similar. A change in the environment may mean that the new animals or plants will not cope with the change and die out.

The gene pool will then contain fewer alleles, reducing further selective breeding options.

In 1996, Dolly the sheep was the first mammal cloned. She died prematurely in 2003. Her early death fuelled the debate about the long-term health problems of clones.

Scientists are looking at ways of using genetically engineered animals to grow replacement organs for humans. This poses many ethical concerns, not least the problems of rejection.

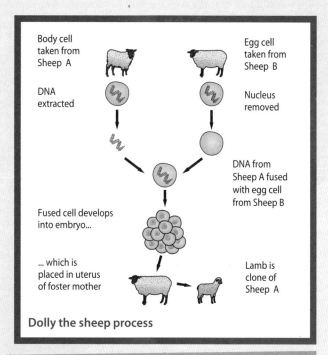

Dolly the sheep process

PROGRESS CHECK

1. What type of reproduction produces clones?

2. Are calves produced by embryo transplant clones?

3. What are the advantages of embryo transplants?

? EXAM QUESTION

Mandy stated that selectively breeding plants was always successful.

a. Is this true or false?

b. What techniques could be used to produce a clone of a plant?

c. Why is selective breeding in animals not always successful?

The Environment

If we look at the information on a food chain and food web, it simply shows us what eats what. A pyramid of numbers, however, shows us how many organisms are involved at each stage in the food chain.

Pyramid of Numbers

At each level of the food chain (trophic level), the number of organisms generally decreases.

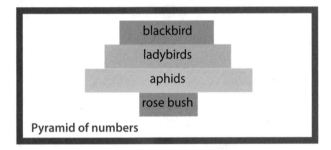

blackbird
ladybirds
aphids
rose bush
Pyramid of numbers

Sometimes a pyramid of numbers does not look like a pyramid at all as the **mass** of the organisms is not taken into account.

Pyramids of Biomass

A biomass pyramid takes into account the mass and numbers of organisms in a food chain.

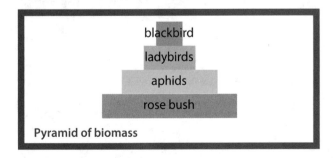

blackbird
ladybirds
aphids
rose bush
Pyramid of biomass

Loss of Energy

Food chains rarely have more than four or five links. The final organism receives only a fraction of the energy that was produced at the beginning of the food chain.

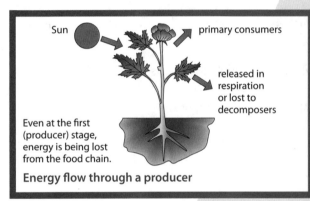

Sun   primary consumers

released in respiration or lost to decomposers

Even at the first (producer) stage, energy is being lost from the food chain.

Energy flow through a producer

Plants absorb their energy from the Sun. Only a small fraction of this energy is converted into glucose during photosynthesis. Some energy is lost to decomposers as plants shed their leaves, seeds or fruit. The plant uses some energy during respiration and growth. The plant's biomass (size) increases and this provides food for the herbivores.

At the primary consumer level, energy losses include waste products and energy used to keep warm.

There are ways to improve the efficiency of food production and reduce energy losses:

■ **Reduce the number of stages in the food chain.** It is more energy efficient to eat plant produce than meat.

- **Intensively rear animals**. By restricting their movement and keeping them warm, animals do not need as much feed. Antibiotics can be used to keep disease at bay.

Photosynthesis

Photosynthesis is a chemical process by which plants make glucose.

The word equation of photosynthesis:

$$\text{carbon dioxide} + \text{water} \xrightarrow[\text{chlorophyll}]{\text{light}} \text{glucose} + \text{oxygen}$$

Uses of Glucose

Glucose is used in **respiration** to obtain energy. It is converted into useful substances such as:

- **insoluble starch** which is stored in the roots, particularly in the winter
- **cellulose** which is needed for cell walls
- **lipids and oils** which are stored in seeds.

Glucose also combines with other substances, such as nitrates from the soil and is turned into **proteins**.

The Rate of Photosynthesis

We can measure the rate of photosynthesis by how much oxygen is produced in a given time. Photosynthesis increases during the summer and plants grow more with increased light and warmth.

There are three things that affect the rate of photosynthesis. We call them **limiting factors**. They are:

- the amount of light
- the amount of carbon dioxide
- temperature.

At any given time, one of these factors could be limiting the rate of photosynthesis.

PROGRESS CHECK

1. How can we reduce energy loss in food chains?

2. What pyramid takes into account the number of organisms in a food chain?

3. What pyramid takes into account the number **and** mass of organisms?

? EXAM QUESTION

1. Complete the word equation of photosynthesis:

$$\underline{\hspace{3cm}} + \text{water} \xrightarrow[\text{chlorophyll}]{\text{light}} \underline{\hspace{2cm}} + \text{oxygen}$$

2. How can we measure the rate of photosynthesis?

3. What **three** factors affect the rate of photosynthesis?

Environmental Damage

An increase in population size leads to an increase in pollution, the destruction of wildlife and greater demands on the world's resources.

Pollution

A pollutant is a substance that harms living things. **Burning fossil fuels** is the main cause of atmospheric pollution. It releases **carbon dioxide** that contributes to the **greenhouse effect**. It also releases **sulfur dioxide** and **nitrogen oxides** that cause **acid rain**. **Methane gas** from cattle and rice fields is another greenhouse gas.

Farming

Farming has become intensive in order to supply more food from a given area of land.

Intensive farming has its problems. Many people regard intensive farming as cruel to animals. In addition, in order to produce more food from the land, fertilisers and pesticides are used.

A possible solution is **organic farming**. Organic farming produces less food per area of land and can be expensive, but it attempts to leave the countryside as it is. It is also kinder to animals. Organic farming uses manure as a **fertiliser, set-aside land** to allow wild plants and animals to flourish, and biological control of pests.

Fertilisers

Plants need nutrients to grow. They take up nutrients from the soil. With intensive farming methods, nutrients are used up quickly, so the farmer has to replace them with artificial fertilisers.

Fertilisers enable farmers to produce more crops from a smaller area of land. This can reduce the need to destroy the countryside for extra space.

There is a problem, however, when fertilisers wash into rivers and lakes. This is called **eutrophication**.

Eutrophication is when plants grow quickly which causes too much competition for light.

As a result, some plants die and are broken down by micro-organisms. The micro-organisms then use up the oxygen in the water and water animals die. Untreated sewage causes the same problem.

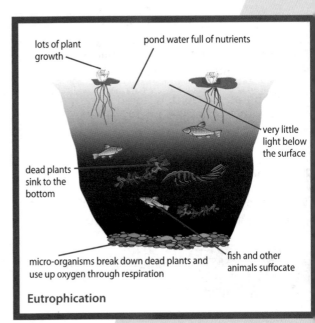

lots of plant growth

pond water full of nutrients

very little light below the surface

dead plants sink to the bottom

micro-organisms break down dead plants and use up oxygen through respiration

fish and other animals suffocate

Eutrophication

Deforestation

In the UK, there are not many forests left. In other countries, forests are being chopped down to provide timber or space for agriculture. This is to try and provide for the growing numbers of people.

The chopping down of forests causes several problems:

- Burning timber increases carbon dioxide in the air.

- Forests absorb carbon dioxide in the air and provide us with oxygen.

- Chopping down trees leads to soil erosion as the soil is exposed to rain and wind.

- Water from trees evaporates into the air and, without it, there will be a decrease in rainfall.

- Destroying forests also destroys many different habitats of animals and plants.

Conservation and Sustainable Development

With the human population increasing and using up resources, we need to find a way of keeping a quality of life for future generations. This is known as **sustainable development**.

It is important to protect our food supply, maintain biodiversity, prevent animals and plants from becoming extinct and conserve resources as we do not know what the future may hold.

PROGRESS CHECK

1. What is the main cause of atmospheric pollution?

2. What gases cause acid rain?

3. Which **two** main gases cause the greenhouse effect?

EXAM QUESTION

1. What is deforestation?

2. Describe the problems deforestation has on wildlife.

3. Why would preserving forests help reduce the greenhouse effect?

Ecology and Classification

We are surrounded by a huge variety of living organisms in a variety of habitats. In any habitat, the numbers of organisms is usually large, so we use sampling techniques to find out what lives there and then estimate the size of the population. There are a variety of techniques used to sample organisms.

Quadrats

A **quadrat** is a wooden or metal frame, usually 1 m² in area. It is placed randomly in a field and the numbers of plants belonging to a species are counted within the quadrat. A plant is counted if more than half of it is inside and touching the quadrat. This is repeated to estimate the number of plants in the whole field.

A quadrat

Pooters and Pitfall Traps

Insects tend to move around, so **pooters and pitfall traps** can be used to collect them, then the insects can be counted and identified.

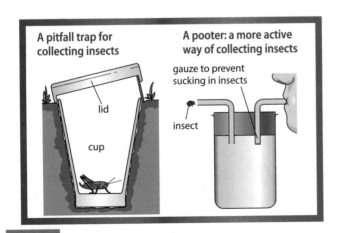

A pitfall trap for collecting insects

A pooter: a more active way of collecting insects

gauze to prevent sucking in insects

insect

lid

cup

Keys

Keys can be used to identify the organisms collected.

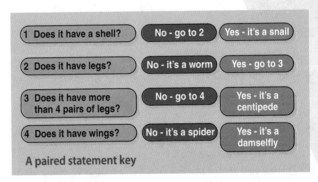

1 Does it have a shell?	No - go to 2	Yes - it's a snail
2 Does it have legs?	No - it's a worm	Yes - go to 3
3 Does it have more than 4 pairs of legs?	No - go to 4	Yes - it's a centipede
4 Does it have wings?	No - it's a spider	Yes - it's a damselfly

A paired statement key

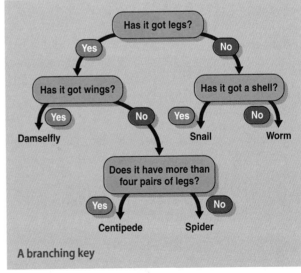

Has it got legs?

Yes — Has it got wings?   No — Has it got a shell?

Yes — Damselfly   No   Yes — Snail   No — Worm

Does it have more than four pairs of legs?

Yes — Centipede   No — Spider

A branching key

Ecosystems

An ecosystem is made up of a community of living things in a habitat. There are **natural ecosystems**, such as woodland or lakes, or **artificial ecosystems** such as greenhouse and aquariums. Artificial ecosystems tend to have a smaller variety of organisms and may use weed killers, pesticides and fertilisers to control growth.

Classification

Classification is sorting living things into groups. Carl Linnaeus devised a classification system based on common features, such as body shape, type of limbs and skeleton. He looked at similarities and differences.

From the largest to the smallest group, the order of classification is:

> Kingdom, Phylum, Class, Order, Family, Genus, Species

A species is a group of living things that are able to breed together to produce fertile offspring. Different species can still be very similar and live in similar types of habitat; they may share a common ancestor.

A species is given a name called a **binomial** consisting of two scientific names. The first shows the genus it belongs to and the second, its species. For example, *Homo sapiens* is the binomial name for humans.

PROGRESS CHECK

1. Who devised the classification system?

2. How many kingdoms exist today?

3. What does 'vertebrate' mean?

Kingdoms

Today scientists recognise five kingdoms.

For many years fungi were classified as plants, but they now belong in their own kingdom as they do not have chlorophyll and cannot make their own food by photosynthesis. To belong to the plant kingdom, organisms must have chloroplasts, cellulose cell walls and make their own food.

To belong in the animal kingdom, organisms have to be able to move, have nerve and muscle cells and must be unable to make their own food.

The animal kingdom is divided into **vertebrates** (animals with backbones) and **invertebrates** (animals without backbones). The vertebrate phylum is divided into five classes that have features in common, e.g. mammals give birth to live young and produce milk.

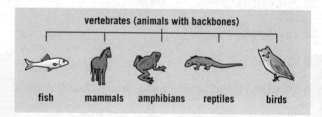

vertebrates (animals with backbones)

fish　　mammals　amphibians　reptiles　　birds

EXAM QUESTION

1. a. What is the definition of a species?

    b. What species does this animal belong to: *Panthera leo?*

2. Joe and David use quadrats to estimate the number of daisies in a field.

    Describe how a quadrat is used to collect reliable data.

Adaptation and Competition

A **habitat** is where an organism lives. A habitat has the conditions needed for it to survive.

A **community** consists of living things in the **habitat**. Each **community** is made up of different **populations of animals and plants** and each **population** is adapted to live in that particular **habitat**.

An **ecosystem** is formed from all the living things and their physical environment. Each **population** is adapted to live in that particular **habitat**.

Sizes of Populations

Population numbers cannot keep growing out of control. Factors that keep the population from becoming too large are called **limiting factors**.

The factors that affect the size of a population are:

- ■ amount of food and water available

- ■ predators or grazing

- ■ disease

- ■ climate, temperature, floods, droughts and storms

- ■ **competition** for space, mates, light, food and water

- ■ human activity, such as pollution or destruction of habitats.

Adaptations

A polar bear would not be found in the desert or a camel in the North Pole. They have **special features** that help them survive in their own environments.

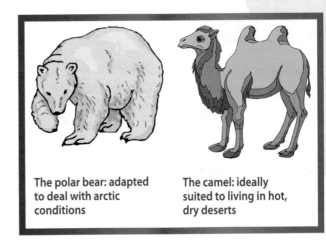

The polar bear: adapted to deal with arctic conditions

The camel: ideally suited to living in hot, dry deserts

A polar bear lives in cold arctic regions of the world; it has many features that enable it to survive.

- ■ It has a **thick coat** to keep in body heat.

- ■ It has a **layer of blubber and fur on the soles of the feet** for insulation.

- ■ It has a **white** coat for camouflage.

- ■ It has **big feet** to spread its weight on snow and ice.

- ■ It has **sharp claws and teeth** to catch seals.

- ■ It has strong legs for **swimming** and **running** to catch prey.

- ■ The shape of its body is **compact** even though it is large. This keeps the surface area to a minimum to reduce body heat.

The cactus has adapted to live in hot, dry conditions.

- To reduce water loss, it has a **rounded shape** (reduced surface area to volume ratio).

- It has **thick cuticles** and leaves reduced to **spines** to also reduce water loss.

- The cactus is capable of **storing water**.

- It has **long roots** to reach water.

- It has a **green stem** for photosynthesis.

In a community, the animal or plant best adapted to its surroundings will survive and be more able to compete for limited resources.

Predator/prey Graphs

In a community, the numbers of animals stays fairly constant. This is partly due to the amount of food limiting the size of the populations.

A predator is an animal that hunts and kills another animal. **A prey** is the hunted animal.

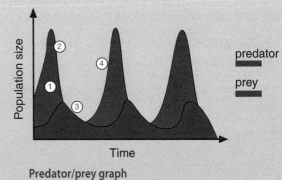

Predator/prey graph

1. An increase in prey means more food for the predator so its numbers increase.

2. The prey then decreases as it is eaten.

3. Predators then decrease, as there is not enough food.

4. Prey numbers can increase again, and so it continues.

Evolution

The **theory of evolution** states that all living things that exist today or existed in the past evolved from simple life forms three billion years ago.

Religious theories are based on the need for a 'creator' of all life on Earth, but there are other theories.

Charles Darwin, a British naturalist, first put forward his ideas about 140 years ago. Darwin visited the Galapagos Islands off the coast of South America and made a number of observations that led to his theory:

1. Organisms produce more offspring than can possibly survive.

2. Population numbers remain fairly constant despite this.

3. All organisms in a species show variation.

4. Some of these variations are inherited.

He concluded from these observations that, since there were more offspring than could survive, there must be a struggle for existence, competition for food, predators and disease. This led to the strongest and fittest offspring surviving and passing on their genes. This is sometimes called the **survival of the fittest** or **natural selection**.

Darwin's theory eventually became widely accepted, but a man called **Lamarck** suggested that animals evolved features according to how much they used them. Giraffes, for example, grew longer necks because they needed to reach food.

Fossils are the remains of dead organisms that lived millions of years ago, found in rocks. Fossils provide evidence for evolution.

The theory of evolution – from animal to human

Natural Selection in Action

An example of the environment causing changes in a species is that of the **peppered moth**. They live in woodlands on lichen-covered trees. There are two types of peppered moth: a light, speckled form and a dark form. The dark form was caused by a mutation and was usually eaten by predators.

In the 1850s, the dark type of moth was rare, but pollution from factories started to blacken tree trunks. The dark moth was then at an advantage because it was camouflaged.

In 1895, most of the population of moths were dark. In cleaner, less polluted areas, the light moth had an advantage against predators and so it still survived to breed.

 dark-coloured moth against a soot-covered tree the pale moth is at a disadvantage in polluted areas

Pollution played a key role in the 'survival of the fittest' for these peppered moths

Extinction

Species that are unable to adapt to their surroundings become extinct. For example, the mammoth, the dodo and the sabre tooth tiger are all examples of extinct species.

Extinction can also be caused by changes in the environment, new predators, new disease, new competition, or human activity, e.g. hunting, pollution or habitat destruction.

An endangered species is a plant or animal that is in danger of becoming extinct, e.g. the panda, gorilla, whale, red squirrel and osprey.

We need to look at ways of protecting habitats, introducing quotas, breeding in captivity, giving endangered species legal protection, education programmes, and even creating artificial ecosystems in which to keep and breed animals.

Extinct species – mammoth, dodo and sabre tooth tiger

PROGRESS CHECK

1. What process causes evolution?

2. What provides some evidence for evolution?

3. What does 'survival of the fittest' mean?

? EXAM QUESTION

The peppered moth exists in two forms, the original light form and a dark form.

a. What process led to the dark form developing?

b. What advantage does the dark moth have over the light moth in city areas?

c. Fill in the gap:

Species that are unable to _____ to their surroundings become extinct.

Air and Air Pollution

Today's atmosphere is composed of: about 78% nitrogen, about 21% oxygen, about 1% argon, small amounts of other gases including carbon dioxide and water vapour. We can obtain nitrogen and oxygen from liquid air by fractional distillation.

Evolution of the Atmosphere

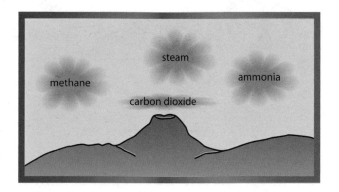

During the first billion years of the Earth's life, there was enormous amounts of volcanic activity. The volcanoes released water vapour, which condensed to form the early **oceans**, while the other gases that were released formed the Earth's early atmosphere.

This atmosphere mainly consisted of **carbon dioxide**, like the present day atmosphere of Mars and Venus.

During the next two billion years, **plants** evolved and began to cover the Earth's surface. These plants steadily removed carbon dioxide and produced oxygen by photosynthesis.

In recent times, people have started to burn large amounts of fossil fuels. This has released large amounts of carbon dioxide back into the atmosphere.

Acid Rain

Many fossil fuels contain small amounts of **sulfur**. When these fuels are burned, this sulfur reacts with oxygen to form sulfur dioxide. If this gas is released into the atmosphere, it can react with rainwater to form acid rain.

We can reduce the amount of acid rain produced by:

■ burning less fossil fuel

■ removing sulfur compounds from oil and natural gas

■ removing sulfur dioxide from the waste gases in coal powered power stations before they are released into the atmosphere.

Global Dimming

Global dimming is caused by smoke particles released into the atmosphere when fuels are burned.

Global Warming

Scientists have found that the temperature of the Earth seems to be gradually increasing.

Carbon dioxide, which is produced when fossil fuels are burned, is thought to be contributing to this effect known as global warming.

The carbon dioxide traps heat energy reaching the Earth from the Sun.

If global warming continues, the polar icecaps may eventually melt. This could cause massive environmental problems.

Nitrogen Oxides

At the high temperatures inside car engines, nitrogen in the atmosphere may react with oxygen to form **nitrogen monoxide**. The nitrogen monoxide is then oxidised to form **nitrogen dioxide**. Nitrogen monoxide and nitrogen dioxide are known as nitrogen oxides or NO_x.

Reducing Pollution from Cars

The amount of atmospheric pollution caused by cars can be reduced by:

- using more efficient engines

- using public transport more

- using catalytic converters; these convert carbon monoxide to carbon dioxide and nitrogen monoxide to nitrogen and oxygen.

PROGRESS CHECK

1. What is the main gas in today's atmosphere?

2. What was the main gas in the Earth's early atmosphere?

3. Why did the amount of carbon dioxide in the Earth's early atmosphere decrease?

4. Name the gas which makes up around 20% of today's atmosphere.

5. Name the gas produced when sulfur reacts with oxygen.

EXAM QUESTION

The Earth's early atmosphere mainly consisted of carbon dioxide.

a. Which planet has an atmosphere similar to the Earth's early atmosphere?

b. Explain why the Earth's atmosphere today has less carbon dioxide and more oxygen than it has had in the past.

c. What is the environmental problem associated with increased levels of carbon dioxide in the atmosphere?

Crude Oil

Crude oil is a mixture. The most important compounds in crude oil are called **hydrocarbons.** Hydrocarbons are compounds that only contain carbon and hydrogen atoms.

Fractional Distillation

Short hydrocarbon molecules are valuable **fuels** as they:

- are easy to ignite

- have low boiling points

- are runny.

Longer hydrocarbon molecules are less valuable, but they are still widely used. Crude oil is separated into **fractions** (groups of molecules with a similar number of carbon atoms) by fractional distillation.

During **fractional distillation** crude oil is heated up until it evaporates. Short hydrocarbon molecules have low boiling points and reach the top of the fractionating column before they condense and are collected.

Longer hydrocarbon molecules have higher boiling points; they condense and are collected lower down the fractionating column.

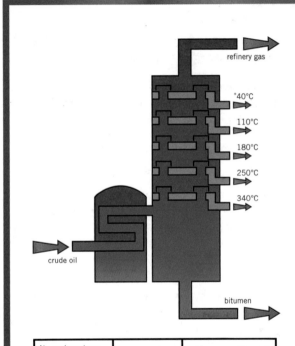

No. carbon atoms in hydrogen chain	Temperature	Fraction collected
3	less than 40°C	refinery gas
8	40°C	petrol
10	110°C	naphtha
15	180°C	kerosene
20	250°C	diesel
35	340°C	oil
50+	above 340°C	bitumen

Fractional distillation column

Uses

Fraction	Use
gases	heating
petrol	fuel for vehicles
naphtha	to make new chemicals
kerosene	jet fuel
diesel and oil	heating and as fuel for vehicles
bitumen	to make roads

Cracking

Fractional distillation of crude oil produces large amounts of long hydrocarbon molecules.

These long molecules can be broken down into smaller, more useful and more valuable molecules by **cracking**. In this process, large molecules are heated until they evaporate and are then passed over a catalyst.

Here decane is cracked to produce octane and ethene:

decane $C_{10}H_{22}$ (from the naphtha fraction)   octane C_8H_{18} ethene C_2H_4

 →

Cracking is an example of a thermal decomposition reaction

PROGRESS CHECK

1. What is a hydrocarbon?

2. What is a 'fraction'?

3. In a fractionating column, where are the shortest hydrocarbon molecules collected?

4. What is the name of the fraction which has the longest hydrocarbon molecules?

5. Which fraction is used for jet fuel?

? EXAM QUESTION

Use the words below to complete the sentences.

condense   **evaporates**
fractional distillation   **fractions**

Crude oil is separated by ____a.____. The crude oil is heated until it ____b.____. Short hydrocarbon molecules reach the top of the column before they ____c.____ and are collected. Groups of molecules with a similar number of carbon atoms are called ____d.____.

Food Additives

Many foods are cooked before we eat them. The cooking of food is a chemical change.

Eggs and meat are good sources of **protein**. Potatoes are a good source of **carbohydrate**.

When we cook eggs or meat, the shape of the protein molecules are changed.

When we cook potatoes we make them easier to digest.

We cook food to make it:

- taste better
- look better
- easier to digest.

The high temperatures used in cooking can also kill micro-organisms in the food.

We sometimes add chemicals to foods to make it:

- look more attractive
- taste better
- have a longer shelf-life.

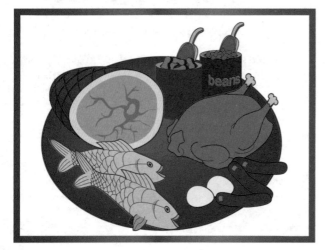

Chemicals that have passed safety tests and are approved for use throughout the European Union, are called **E-numbers**.

Food additives can be natural substances or artificial substances.

Chemicals commonly added to foods include:

- **emulsifiers** – used to keep unblendable liquids mixed together
- **colours** – used to make food look more attractive (artificial colourings can be detected by chromatography)
- **flavours** – used to make the food taste better
- **artificial sweeteners** – used to reduce the amount of sugar needed
- **preservatives** – these increase the shelf-life of the food by stopping harmful micro-organisms from growing
- **antioxidants** – increase the shelf-life of the foods that contain oils and fats by preventing reactions with oxygen.

Public Concern

The number of additives in food has caused some public concern. Scientists believe, however, that provided we eat a balanced diet including lots of different foods, the permitted food additives are safe for the vast majority of people.

Baking Powder

Baking powder is used to make biscuits and cakes.
It contains **sodium hydrogen carbonate**, $NaHCO_3$.

When sodium hydrogen carbonate is heated,
a thermal decomposition reaction takes place
producing sodium carbonate, water and carbon
dioxide.

sodium hydrogen carbonate   →   sodium carbonate + water + carbon dioxide

$$2NaHCO_3 \quad \rightarrow \quad Na_2CO_3 \quad + \quad H_2O \quad + \quad CO_2$$

The products include carbon dioxide gas which
becomes trapped in the dough mixture and makes
the dough rise. This gives cakes and biscuits a light,
pleasing texture.

PROGRESS CHECK

1. Where are E-numbers approved for use?

2. What does an emulsifier do?

3. Is a strawberry an artificial or a natural substance?

4. What type of additive can be added to a food to make it look better?

5. What type of food additive is added to oils and fats to stop them from reacting with oxygen?

EXAM QUESTION

1. What is the name of the alkali found in baking powder?

2. Why is baking power used when we make cakes or biscuits?

3. Complete the word equation to show the products of the thermal decomposition of sodium hydrogen carbonate:

   sodium hydrogen carbonate → sodium carbonate + water + _____

Vegetable Oils

Vegetable oils are important foods. They are a good source of **energy** and of the vitamins A and D.

Oils as Foods

Popular vegetable oils include sunflower oil and olive oil.

Vegetable oils can also be used as **biofuels**. When they are burned, they release lots of energy.

Vegetable oils can be produced from:

- **seeds**
- **nuts**
- **fruits**.

First the plant material is crushed, then the oil is extracted.

Vegetable oils can be extracted from the fruits, seeds and nuts of some plants

Vegetable oils are unsaturated because they contain carbon double bonds. We can check this by adding bromine water. The bromine water changes colour from brown to colourless.

Some vegetable oils contain many carbon double bonds. These are described as **polyunsaturated** fats. Doctors believe that polyunsaturated fats are good for our health.

Emulsions

An **emulsion** is a mixture of two immiscible (do not blend together) liquids.

Salad dressing is an example of an everyday emulsion. Salad dressings are made by shaking oil with vinegar so that the two liquids mix together, but the oil and vinegar soon separate out.

To stop this from happening we can add an **emulsifier**. One end of an emulsifier molecule is attracted to the water in vinegar (hydrophilic) while the other end is attracted to the oil molecules (hydrophobic).

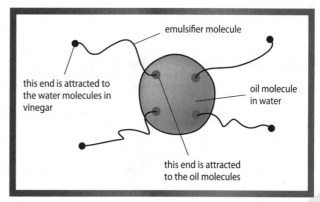

emulsifier molecule

this end is attracted to the water molecules in vinegar

oil molecule in water

this end is attracted to the oil molecules

Hydrogenated Vegetable Oils

Most vegetable oils are liquids at room temperature. Liquid oils can be very useful, but there are times when we might prefer a solid fat, for example, when we want to spread it onto bread or make cakes.

Vegetable oils can be made solid at room temperature by a process known as hydrogenation. The vegetable oils are heated with hydrogen using a nickel catalyst at around 60°C.

Paint

Paints can be used to make things look more attractive, or to protect them. Paint is a special type of mixture called a **colloid**. It consists of tiny coloured particles which are suspended in a solvent. Types of paint include oil paints and emulsion paints.

PROGRESS CHECK

1. What vitamins are found in vegetable oils?

2. Name **two** types of vegetable oil.

3. Which parts of plants can we extract vegetable oils from?

4. Why can vegetable oils be described as 'unsaturated'?

5. What do salad dressings contain?

EXAM QUESTION

Use the words below to complete the table:

emulsion   bromine water   biofuel   polyunsaturated

Name	Description
a.	A mixture of two immiscible liquids
b.	A fat with many carbon double bonds
c.	A material made from living organisms that can be burned to release energy
d.	A chemical used to test for a carbon double bond

Fuels

A fuel is a substance which can be burned to release energy.

This means the burning of a fuel is an exothermic reaction.

A good fuel should:

- be easy to ignite
- be widely available
- release lots of energy when it is burned
- burn cleanly without producing lots of soot
- not damage the environment.

Most fuels contain carbon and/or hydrogen.

Hydrocarbons are compounds which only contain carbon and hydrogen atoms. Petrol, diesel and fuel oil are all hydrocarbons. Coal is mainly made of carbon.

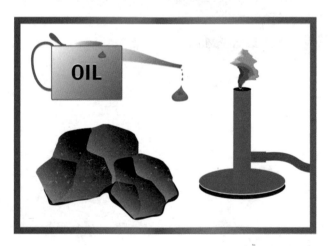

Combustion

Complete combustion of hydrocarbon fuels produces **carbon dioxide, CO_2** and **water vapour, H_2O**.

Scientists believe that carbon dioxide contributes towards global warming.

Incomplete combustion of hydrocarbon fuels can occur if there is a limited supply of oxygen. This can occur in faulty gas appliances. The products of incomplete combustion include **carbon monoxide, CO** and carbon.

Carbon monoxide is a toxic gas. It lowers the ability of red blood cells to carry oxygen around the body.

Unburnt carbon makes a flame look yellow and it is deposited as **soot** making surfaces dirty.

Hydrogen

Hydrogen is used as a fuel in space rockets, but currently it is not used as a fuel for cars.

Advantages of using hydrogen as a fuel	Disadvantages of using hydrogen as a fuel
Releases lots of energy when it is burned	It is a gas which can only be liquefied at very low temperatures or very high pressures
Burns very cleanly	It is difficult to store – a car would require a large, heavy container to store the hydrogen
The only product of combustion is water vapour which does not harm the environment	Hydrogen is not found naturally and would have to be made before it could be used. This would probably require the use of fossil fuels

Biofuels

Biofuels are fuels which are produced from living things. Wood and ethanol are examples of biofuels.

PROGRESS CHECK

1. Which **two** elements are found in hydrocarbon fuels?

2. What are the products of the complete combustion of hydrocarbon fuels?

3. Name the main element found in coal.

4. What is the environmental problem associated with carbon dioxide?

5. Why is the incomplete combustion of hydrocarbon fuels undesirable?

? EXAM QUESTION

1. Complete the word equation below to show the product of the complete combustion of carbon.

   carbon + oxygen → _____

2. Carbon can also be burned in a limited supply of oxygen. Name **one** of the products of the incomplete combustion of carbon.

3. Many fuels also contain atoms of hydrogen. Complete the word equation to show the product of the combustion of hydrogen.

   hydrogen + oxygen → _____

Alkanes and Alkenes

Carbon atoms form four bonds with other atoms while hydrogen atoms only form one bond.

Alkanes

The **alkanes** are a family of **hydrocarbon** molecules. Hydrocarbon molecules only contain carbon and hydrogen atoms. Alkanes are saturated hydrocarbons because they do not contain carbon double bonds. Short alkane molecules are useful fuels.

Alkanes have the general formula C_nH_{2n+2}.

Name	methane	ethane	propane	butane
Chemical formula	CH_4	C_2H_6	C_3H_8	C_4H_{10}
Structure	H | H–C–H | H	H H | | H–C–C–H | | H H	H H H | | | H–C–C–C–H | | | H H H	H H H H | | | | H–C–C–C–C–H | | | | H H H H

The lines in these structural diagrams represent covalent bonds

Alkenes

The **alkenes** are another family of hydrocarbon molecules.

Alkenes are produced during cracking.

Alkenes are unsaturated hydrocarbons because they do contain carbon double bonds. Alkene molecules are more reactive than alkane molecules.

They can be used to make new chemicals including plastics.

Alkenes have the general formula C_nH_{2n}.

Name	ethene	propene
Chemical formula	C_2H_4	C_3H_6
Structure	H C=C H H H	H H H C=C–C–H

Polymerisation

In addition **polymerisation**, many small molecules are joined together to form a bigger molecule. The small molecules are called **monomers**. The bigger molecule is called a polymer. Plastics are **polymers**.

The diagram below represents the reaction between lots of ethene molecules to form polythene.

$$n \; \begin{array}{c} H \;\; H \\ C=C \\ H \;\; H \end{array} \rightarrow \left(\begin{array}{c} H \;\; H \\ C–C \\ H \;\; H \end{array} \right)_n$$

The diagram below represents the reaction between lots of propene molecules to form polypropene.

$$n \ \underset{\underset{H}{|}}{\overset{\overset{CH_3}{|}}{C}} = \underset{\underset{H}{|}}{\overset{\overset{H}{|}}{C}} \rightarrow \left(\underset{\underset{H}{|}}{\overset{\overset{CH_3}{|}}{C}} - \underset{\underset{H}{|}}{\overset{\overset{H}{|}}{C}} \right)_n$$

Properties of Plastics

The structure and bonding within a material affects its properties. The stronger the forces between the particles in a solid the higher the temperature at which the solid melts.

Modifying the structure of a polymer can influence its properties.

■ Adding a plasticiser can make a plastic more flexible.

■ Lengthening the polymer chain increases the forces of attraction between the molecules and increases the melting point of the polymer.

■ Making the polymer chains more aligned increases the forces of attraction between the molecules and increases the melting point.

■ Some plastics consist of long polymer chains with very little cross linking between the chains. When these plastics are heated the chains untangle and the plastic softens. This means that these plastics can be reshaped many times.

■ Other plastics consist of polymer chains that are heavily cross linked. These polymers must be shaped when they are first made. When they are heated again they will not soften but may eventually burn.

PROGRESS CHECK

1. What is the general formula of an alkane?

2. Give a use for a short alkane molecule.

3. What do the lines in structural diagrams represent?

4. How can alkenes be made?

5. What is the formula of ethene?

EXAM QUESTION

Polythene is a polymer.

a. Name the monomer used to make polythene.

b. Which hydrocarbon family does the monomer used to make polythene come from?

c. What is the general formula of this hydrocarbon family?

Limestone

Limestone is a type of sedimentary rock. It is mainly made of **calcium carbonate**, $CaCO_3$.

When limestone is heated it decomposes to form calcium oxide and carbon dioxide:

> calcium carbonate → calcium oxide + carbon dioxide

This is an example of a **thermal decomposition** reaction.

Calcium oxide is also known as **quicklime**. Quicklime can be reacted with water to form calcium hydroxide:

> calcium oxide + water → calcium hydroxide

Calcium hydroxide is also known as **slaked lime**.

Other metal carbonates react in a similar way when they are heated.

More Useful Materials

Other useful materials made from limestone include:

- cement – made by heating powdered limestone and powdered clay and then adding water
- mortar – made by mixing cement, sand and water
- concrete – made by mixing cement, sand, rock chippings and water
- glass – made by heating up a mixture of limestone, sand and soda until it melts.

Reinforced concrete is a composite material made by allowing concrete to set around steel supports.

Clues in Names

The name calcium carbon**ate** tells us the compound contains calcium, carbon and lots of oxygen.

The names of some other compounds such as potassium nit**rite** have a slightly different ending.

If a compound ends in 'ite' it contains some oxygen but not as much as if the name of the compound ended in 'ate'.

Potassium nitrite, therefore, must contain potassium, nitrogen and some oxygen.

Making Salts

Salts are very important compounds. They can be made by reacting acids with metal carbonates, metal oxides or metal hydroxides.

Sulfuric acid forms sulfate salts while hydrochloric acid forms chloride salts.

Metal Carbonates

Metal carbonates react with acids to form a salt, water and carbon dioxide.

Metal Oxides

Metal oxides react with acids to form a salt and water.

Metal Hydroxide

Metal hydroxides also react with acids to form a salt and water.

Other Rocks

Granite is an example of an igneous rock. Igneous rocks are formed when molten rock cools down and solidifies.

If the rock formed quickly, it will contain small crystals for example basalt.

If the rock formed slowly, it will contain large crystals for example gabbro.

If limestone is subjected to high temperatures or pressures it can be turned into the metamorphic rock marble.

PROGRESS CHECK

1. Name the main chemical compound in limestone.

2. What type of rock is limestone?

3. Name the products of the thermal decomposition of limestone.

4. By what name is calcium oxide also known?

5. By what name is calcium hydroxide also known?

EXAM QUESTION

1. What type of rock is granite?

2. Name the rock formed when limestone is subjected to high temperatures and pressure.

Cosmetics

A solution is a mixture made when a solvent dissolves a solute. Water is a good solvent for many, but not all, substances. Some solutes like nail varnish are insoluble in water.

Perfumes

Ethanol is often used as a solvent in perfumes. Traditional perfumes contain plant and animal extracts such as jasmine and lavender, but these ingredients can be very expensive.

Today we often use cheaper, man-made fragrances such as esters. Esters are made by reacting carboxylic acids with alcohols.

A good perfume should:

- evaporate easily from the skin

- be non-toxic

- not react with water (sweat)

- not irritate the skin

- be insoluble in water so it is not washed off easily.

Perfumes evaporate because although there are strong forces of attraction within perfume molecules, there are much weaker forces of attraction between perfume molecules. When the perfume is put on the skin, some of the molecules gain enough energy to evaporate.

Animal Testing

Cosmetic products have to be tested before they can be sold. Some tests involve living animals. Some people believe that this causes avoidable suffering to animals, while other people think that animal testing is the best way to ensure that products are safe for people to use.

Making Ethanol

Ethanol is a type of alcohol. It has the structure:

$$H-\overset{\displaystyle H}{\underset{\displaystyle H}{C}}-\overset{\displaystyle H}{\underset{\displaystyle H}{C}}-O-H$$

Ethanol can be made from sugar cane or sugar beet. It is a useful biofuel (a fuel made from living materials). Unlike petrol, ethanol is a renewable fuel but large areas of fertile land must be used to grow the sugar cane or sugar beet.

Ethanol is produced by **fermentation**. During fermentation, yeast converts glucose (sugar) into ethanol:

$$\text{glucose} \xrightarrow{\text{yeast}} \text{ethanol} + \text{carbon dioxide}$$

$$C_6H_{12}O_6 \longrightarrow 2C_2H_5OH + 2CO_2$$

Fermentation

Fermentation has been used to produce alcoholic drinks such as wine and beer. The consumption of alcoholic drinks causes many social and health problems and is banned by some religions.

Ethanol can also be made from non-renewable sources. Ethanol produced in this way is called industrial alcohol. Ethene, which is produced by the cracking of long chain hydrocarbons, is reacted with steam to produce ethanol.

ethene + steam → ethanol

$$C_2H_4 + H_2O \rightarrow C_2H_5OH$$

PROGRESS CHECK

1. What is a solution?

2. Name a solute which is insoluble in water.

3. Name a solvent often used in perfumes.

4. Write down a plant extract used in perfumes.

5. How can an ester be made?

EXAM QUESTION

Ethanol is a type of alcohol.

a. What is the name of the reaction in which glucose is converted into alcohol and carbon dioxide?

b. Ethanol can also be made by reacting ethene with steam. Write a word equation for this reaction.

Structure of the Earth

The Earth has a layered structure.

- At the centre of the Earth is the **core**. The core is divided into two parts: the solid inner core and the outer core which is liquid.

- The core is surrounded by the **mantle**.

- Around the mantle is the **crust**.

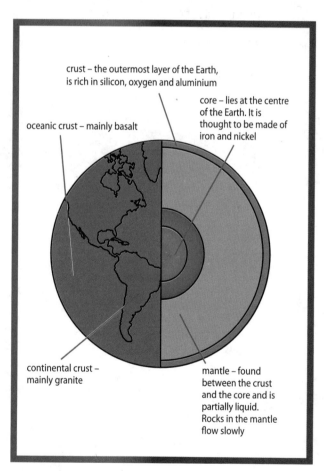

crust – the outermost layer of the Earth, is rich in silicon, oxygen and aluminium

core – lies at the centre of the Earth. It is thought to be made of iron and nickel

oceanic crust – mainly basalt

continental crust – mainly granite

mantle – found between the crust and the core and is partially liquid. Rocks in the mantle flow slowly

Plate Tectonics

People used to believe that the features that we see at the Earth's surface were formed as the Earth cooled down. Scientists now believe that these features were caused by **plate tectonics**.

In this theory, the Earth's **lithosphere** (crust and the upper part of the mantle) is broken into about a dozen pieces called plates. These plates are carried by convection currents in the Earth's mantle.

The currents are caused by the heat released by natural radioactive decay. The plates move a few centimetres each year. This is about the same rate as your finger nails grow.

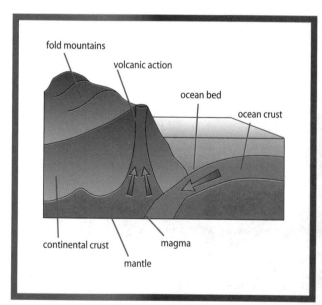

fold mountains

volcanic action

ocean bed

ocean crust

continental crust

magma

mantle

Earthquakes

Sometimes the plates cannot move smoothly. **Earthquakes** occur where the plates try to move past each other but become stuck. The forces on these plates gradually build up. Eventually the plates move and they release the strain that has built up as an earthquake, or if it happens under water, a **tsunami**.

Volcanoes

Volcanoes are also found at plate boundaries. When an oceanic plate collides with a continental plate, the denser oceanic plate is subducted beneath the continental plate. Some of the oceanic plate may melt to form magma (molten rock which is underground).

The hot magma may be less dense than the surrounding rock; if it is, it may rise to the surface through cracks to form volcanoes. Volcanoes can cause devastating loss of life.

If the magma is iron-rich, it will be quite runny and the volcano will erupt relatively slowly and safely.

If the magma is silica-rich, however, it will be much more viscous. The volcano will erupt explosively producing volcanic ash and throwing out molten rock called 'bombs'. These volcanoes are very dangerous to local people.

Although our methods of predicting when earthquakes and volcanoes will happen are improving, we still cannot say exactly when they will occur.

PROGRESS CHECK

1. In what state is the inner core?

2. In what state is the outer core?

3. What is the name of the layer between the outer core and the crust?

4. Roughly how many plates are there?

5. What can form if an earthquake happens under the sea?

EXAM QUESTION

The Earth's lithosphere is split into about a dozen plates.

a. What is the lithosphere?

b. At what rate do these plates move?

c. Historically, how did people believe that mountains formed?

New Materials

Gore-tex™ is a new material used to make waterproof objects like jackets. It consists of a thin membrane which is used to coat fabrics.

The membrane has lots of little holes. Liquid water is too big to go through these holes, so the fabric is waterproof. Water vapour is small enough to pass through the holes so it is **breathable**.

In the past, waterproof jackets were made from nylon. This material was lightweight, hard wearing and waterproof but because it was not breathable, perspiration could make a nylon coat quite uncomfortable to wear.

A Gore-tex™ jacket

Kevlar

Kevlar is:

- lightweight

- very strong

- flexible.

Kevlar is used to make bulletproof vests.

Lycra

Lycra is a very **stretchy** material. It can be mixed with other fibres to make fabrics which can be used to make swimsuits.

Thinsulate

Thinsulate is an insulating material. It contains very small **fibres**. These fibres trap air which acts as a layer of insulation stopping body heat from escaping, so it keeps you warm.

PTFE

PTFE is better known by its trade name of 'Teflon'. Its properties are as follows:

- It is very unreactive.

- It has a very slippery surface.

PTFE was first discovered accidentally by scientists investigating refrigerant gases. Today it is used to make non-stick saucepans.

Carbon Fibres

Carbon fibres:

- have a low density

- do not stretch

- are not compressed.

Carbon fibres are mixed with epoxy to make sports equipment such as squash racquets.

Smart Materials

Smart materials are very special materials which have one or more **property** that can be dramatically modified by changes in the environment.

Thermochromic pigments can be added to paints. When these paints are warmed up, the colour of the paint changes. Applications include designs on cups that appear when hot liquids are added.

Nitinol is another smart material. When a force is applied, nitinol stretches but when it is warmed up, it returns to its original shape.

Phosphorescent Pigments

These pigments absorb light energy and then release it over a period of time. They can be used to make objects that glow in the dark such as wristwatch dials.

Nanoparticles

Nanoparticles consist of just a few hundred atoms, so they are incredibly small. Present uses of these materials include sunscreens. Future uses could include better, smaller computers.

PROGRESS CHECK

1. Why can liquid water not pass through Gore-tex™?

2. Which material was widely used to make waterproof jackets in the past?

3. What is the trade name for PTFE?

4. How is PTFE used?

5. What is Kevlar used to make?

EXAM QUESTION

Thinsulate is a very useful material.

a. Describe the structure of thinsulate.

b. Explain why thinsulate might be used to make a hat.

Pollution

Limestone is an important raw material. It can be used as a building material and can be made into other important materials like cement and glass.

Limestone

The advantages of limestone quarrying include the economic and social benefits of new jobs for local people.

The disadvantages of a limestone quarry include noise and dust from the quarry and problems created by lorries transporting the limestone.

Plastics

Plastics are a type of **polymer**. The properties of plastics depend on how they are made and what they are made from.

Polymers are very widely used and new applications are being developed including new **'intelligent' packaging**. 'Intelligent' packaging is used to improve the quality and safety of foods, for example, they can be used to remove water from inside a packet so that it is more difficult for bacteria or mould to grow and the food will stay fresh for longer.

Problems with Plastics

Plastics are very useful. Most plastics, however, are **non-biodegradable**. This means that when plastic objects are thrown away, they remain in the environment. This can cause problems:

- Landfill sites fill up more quickly.

- If we try to get rid of the plastics by burning them, toxic gases may be produced.

- It is very difficult to recycle plastics because it is hard to separate them out.

Because of these problems, scientists have developed a range of plastics which are biodegradable.

Recycling

Recycling waste materials including glass, metal and paper helps to protect the environment because fewer raw materials are required. This also means that less waste needs to be disposed of.

Sustainable Development

Sustainable development balances the need for **economic development** with a respect for the **environment**, so that people can enjoy a good standard of living today without compromising the needs of future generations.

PROGRESS CHECK

1. How can a new limestone quarry benefit people?

2. What are the disadvantages of having a limestone quarry nearby?

3. Name a type of polymer.

4. What could be a disadvantage of burning plastics?

5. Why is it difficult to recycle plastics?

? EXAM QUESTION

Scientists have developed 'intelligent' packaging for food.

a. What can 'intelligent' packaging do?

b. Why is this an advantage to consumers?

Useful Metals

Pure aluminium has a low **density** but is too soft for many uses.

Aluminium can be mixed with other metals to form an **alloy**. These alloys combine low density with high strength.

Other common alloys are:

- amalgram (mainly mercury)
- **brass** (copper and zinc)
- solder (lead and tin)
- bronze (copper and tin).

Aluminium is a reactive metal. It is extracted from its ore, bauxite, by **electrolysis**. This is an expensive process because it involves many steps and requires lots of energy.

Aluminium is actually much more reactive than it appears. This is because aluminium objects quickly react with oxygen to form a layer of aluminium oxide, which prevents any further reaction from occurring.

Aluminium can be used to make many things including:

- drinks cans
- bicycles
- aeroplanes.

Aluminium Cars

Car bodies are normally made from steel, but they can also be made from aluminium.

The advantages of using aluminium are as follows:

- The car body will be lighter so the car will have a better fuel economy.
- The car body will corrode less so it may last for longer.

One of the disadvantages of using aluminium is that an aluminium car body will be more expensive to produce.

Protection of the Environment

The advantages of **recycling** old aluminium objects are as follows:

- Landfill sites are not filled up as quickly.
- Recycling old aluminium objects uses much less energy than extracting aluminium from its ore.
- Recycling old aluminium objects means that less aluminium ore needs to be extracted which protects the environment.

Titanium

Titanium is a very useful metal.

- It has a low density.
- It is strong.
- It is very resistant to corrosion.
- It has a very high melting point.

Titanium is a reactive metal, so extracting it from its ore, rutile, is a difficult process.

Copper

Copper is a very useful metal, which is:

- a good electrical conductor
- a good thermal conductor
- very resistant to corrosion
- very unreactive.

Copper is widely used for electrical wiring and plumbing. Traditionally, we extract copper from its ores and then purify it using electrolysis. Today we have to extract copper from ores which actually contain very little copper. This means that very large quantities of rock must be quarried which cause environmental problems. Scientists are developing ways to extract copper from low grade ores to try to minimise these environmental problems.

PROGRESS CHECK

1. Why is pure aluminium not widely used?

2. What is a mixture of metals called?

3. What is brass made from?

4. Give **one** use of aluminium.

5. What is the ore of titanium called?

? EXAM QUESTION

Aluminium can be used to make car bodies.

a. Why does aluminium appear to be less reactive than it really is?

b. Give **one** advantage of using aluminium rather than steel to make car bodies.

c. Give **one** disadvantage of using aluminium rather than steel to make car bodies.

Iron and Steel

Gold is a very unreactive metal. It is found in nature uncombined. Iron is a fairly reactive metal. It is extracted from its ore which contains iron (III) oxide in the **blast furnace**.

An **ore** is a rock which contains metals in such high concentrations that it is economically worthwhile to extract the metal from the rock.

Iron is less reactive than carbon. To extract iron from iron (III) oxide, the oxygen must be removed. This reaction is called reduction. Iron produced by the blast furnace contains quite large amounts of carbon. If it is allowed to cool down and solidify, **cast iron** is produced. Cast iron contains around 96% pure iron.

Cast iron is:

- very hard
- strong
- resistant to corrosion
- very brittle.

Wrought Iron

Wrought iron is produced by removing the carbon from cast iron. Because it is very pure the iron atoms form a very regular arrangement. This means that the layers of iron atoms can slip over each other very easily, which makes wrought iron very easy to shape. Unfortunately, it can be too soft for many uses.

Steel

Most iron is made into **steel**. First the carbon impurities are removed, then carefully controlled amounts of carbon and metals like chromium are added.

We can produce steels which have quite different properties.

- Low carbon steels are soft and easy to shape.
- High carbon steels are hard but brittle.
- **Stainless steel** is very resistant to corrosion.

Steel is harder than wrought iron because it contains atoms of iron, carbon and other metals which are different sizes. This means the atoms cannot form a regular arrangement, which makes it very difficult for the layers to slip over each other. This is what makes steel hard.

Steels are also less likely to corrode than pure iron.

Rusting

The corrosion of iron is called rusting. During rusting, iron reacts with water and oxygen to form hydrated iron (III) oxide.

This can be summarised by the word equation:

iron + water + oxygen → hydrated iron (III) oxide

Salt increases the rate of rusting, so cars at the seaside may rust faster than cars away from the coast.

PROGRESS CHECK

1. Name the iron compound found in iron ore.

2. Name a metal which is found uncombined in nature.

3. Where is iron extracted?

4. Iron is made by removing oxygen from iron oxide. What is this type of reaction called?

5. Roughly, what is the percentage of iron in cast iron?

? EXAM QUESTION

Use the words below to complete the table.

cast iron   **wrought iron**   **low carbon steel**   **iron (III) oxide**

Name	Description
a.	The name of very pure iron
b.	An alloy containing iron and carbon which is easy to shape
c.	A material produced when iron from the blast furnace is allowed to cool down and solidify
d.	The compound found in iron ore

Salts

Salts are very important compounds.

Uses of salts include:

- the production of fertilisers
- as colouring agents
- in fireworks.

Insoluble Salts

Some salts are **insoluble**. They can be made by reacting solutions of soluble salts.

Here insoluble barium sulfate is made by reacting a solution of barium chloride with a solution of sodium sulfate.

The reaction also produces sodium chloride.

The barium sulfate is a **precipitate**.

| barium chloride | + | sodium sulfate | → | barium sulfate | + | sodium chloride |
| $BaCl_2$ (aq) | + | Na_2SO_4 (aq) | → | $BaSO_4$ (s) | + | $2NaCl$ (aq) |

The state symbol (aq) means aqueous (it is dissolved in water).
The state symbol (s) means it is solid.

A pure sample of the insoluble barium sulfate is obtained by filtering, washing and then drying the salt.

Sodium Chloride

Sodium chloride (common salt) is a very important resource. It is found in large quantities dissolved in seawater and in underground deposits formed when ancient seas evaporated.

Rock salt (unpurified salt) is used in winter to grit roads to stop them from becoming icy and dangerous.

The Electrolysis of Sodium Chloride

The **electrolysis** of concentrated sodium chloride solution produces chlorine, hydrogen and sodium.

- The chlorine is produced at the positive electrode.
- Hydrogen is produced at the negative electrode.
- Chlorine is used to sterilise water.
- Hydrogen is used in the manufacture of margarine.
- Sodium hydroxide is used in the production of soaps.

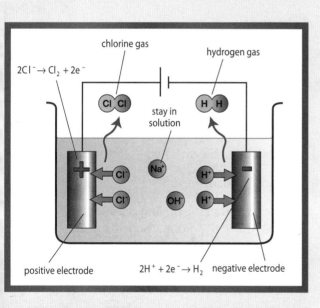

chlorine gas

hydrogen gas

$2Cl^- \rightarrow Cl_2 + 2e^-$

stay in solution

positive electrode

$2H^+ + 2e^- \rightarrow H_2$ negative electrode

The electrolysis of molten sodium chloride produces sodium and chlorine.

PROGRESS CHECK

1. Give a use of salts.

2. Name an insoluble salt.

3. Give the word equation for the reaction between barium chloride and sodium sulfate.

4. What does the state symbol (s) indicate?

5. What does the state symbol (aq) indicate?

? EXAM QUESTION

Sodium chloride is an important resource. Unpurified sodium chloride is called rock salt.

a. How is rock salt used?

b. The electrolysis of concentrated sodium chloride solution produces two gases. Chlorine is produced at the positive electrode. Which gas is produced at the negative electrode?

c. Give one use of chlorine.

The Periodic Table

The noble gases include: helium, neon, and argon.

They are found on the far right-hand side of the **periodic table**. All the noble gases are very **unreactive** but this can make them very useful.

Helium is **less dense** than air, so if a balloon is filled with helium gas, it will float. Helium balloons are popular decorations at parties and special events.

Argon is also a noble gas. It is used to make filament lamps.

Neon is widely used in electrical discharge tubes.

Alkali Metals

The alkali metals include:

- lithium
- sodium
- potassium.

They are found on the far left-hand side of the periodic table. All the alkali metals are very **reactive.** They react with water to form a metal hydroxide and hydrogen.

Example

sodium + water → sodium hydroxide + hydrogen

There is a gradual increase in reactivity down the group.

Halogens

The halogens include:

- chlorine
- bromine
- iodine.

They are found next to the noble gases in the periodic table.

- Chlorine is a pale green gas.
- Bromine is a brown liquid.
- Iodine is a dark grey solid.

This shows us that the **boiling point** of the halogens increases down the group. There is a gradual decrease in reactivity down the group, so chlorine is more reactive than bromine and iodine.

A more reactive halogen will displace a less reactive halogen from its solution. So chlorine will displace bromine from a solution of potassium bromide.

Example

chlorine + potassium bromide → potassium chloride + bromine

Transition Metals

The transition metals include:

- copper
- iron
- gold
- silver.

They are found in the middle of the periodic table. Transition metals:

- are good thermal conductors
- are electrical conductors
- can be bent or hammered into shape.

Gold and silver are used to make jewellery.

PROGRESS CHECK

1. Which group does chlorine belong to?
2. Which group does neon belong to?
3. Which group does lithium belong to?
4. How is helium used?
5. How is argon used?

? EXAM QUESTION

Chlorine and bromine both belong to the same group of the periodic table.

a. Which group do they belong to?

b. Chlorine is more reactive than bromine. Write a word equation to sum up the reaction between chlorine and potassium bromide.

Atomic Structure

Elements are made of only one type of **atom**.

Atoms consist of:

- a small, central **nucleus** (which contains protons and neutrons)
- surrounded by shells of electrons.

All atoms of the same element have the same number of protons.

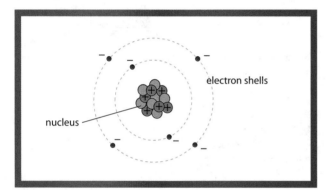

Bonding

Atoms can join together by:

- sharing electrons (this is called a covalent bond)
- transferring electrons (this forms ions, the attraction between oppositely charged ions is called an ionic bond).

The Periodic Table

- There are about 100 different types of element.
- These elements are often displayed in the periodic table.
- The horizontal rows are called **periods**.
- The vertical columns are called **groups**.

Group 1 of the periodic table is sometimes known as the alkali metals.

Group 7 is known as the halogens.

Group 0 is known as the noble gases.

When the periodic table was first designed, many of the elements which we know today had yet to be discovered. Gaps were left in the table and detailed predictions were made about what the new elements would be like.

Symbols

Elements can be represented using symbols.

Symbols are one or two letter codes:

- O is used for oxygen
- C is used for carbon
- Ca is used for calcium
- K is used for potassium
- Na is used for sodium
- Pb is used for lead
- Fe is used for iron

Formula

The chemical formula tells us the number and type of atoms present.

- An oxygen molecule has the formula O_2. This tells us that an oxygen molecule consists of two oxygen atoms.

- A nitrogen molecule has the formula N_2. This tells us that a nitrogen molecule consists of two nitrogen atoms.

- Carbon dioxide has the formula CO_2. This tells us that a carbon dioxide molecule consists of one carbon atom and two oxygen atoms.

- Calcium hydroxide has the formula $Ca(OH)_2$. This tells us that the ratio of the atoms is one calcium atom: two oxygen atoms: two hydrogen atoms.

- Water has the formula H_2O. This tells us that a water molecule consists of two hydrogen atoms and one oxygen atom.

- Copper sulfate has the formula $CuSO_4$. This tells us that the ratio of the atoms is one copper atom: one sulfur atom: four oxygen atoms.

Equations

We can use word equations to sum up what happens during a chemical reaction.

For example, when carbon is burned in plenty of oxygen, carbon dioxide is produced. This can be written as:

$$carbon + oxygen \rightarrow carbon\ dioxide$$

This reaction can also be summed up using the symbol equation:

$$C + O_2 \rightarrow CO_2$$

When hydrogen is burnt in oxygen water vapour is produced. This can be written as:

$$hydrogen + oxygen \rightarrow water\ vapour$$

This can also be written as:

$$H_2 + \tfrac{1}{2} O_2 \rightarrow H_2O$$

During a chemical reaction, atoms are not created or destroyed, they are simply rearranged. This means that there must be equal numbers of each type of atom on both sides of the equation.

Chemicals

Different chemicals have different uses.

Chemical	Use
Ammonia	To make fertilisers
Carbohydrate	A type of food we need for energy
Carbon dioxide	To make carbonated drinks
Caustic soda	To make bleach
Citric acid	To flavour foods and drinks
Ethanoic (acetic) acid	In vinegar
Hydrochloric acid	To remove limescale from boilers
Phosphoric acid	A catalyst for the production of industrial alcohol
Sodium chloride	To 'grit' roads in winter
Water	In industry and in agriculture

Gas Tests

Hydrogen
Burns with a 'squeaky' pop.

Oxygen
Relights a glowing splint.

Carbon Dioxide
When bubbled through limewater, it turns the limewater cloudy.

Ammonia
Turns damp red litmus paper blue.

Chlorine
Bleaches damp litmus paper.

Collecting Gases

- Upward delivery is used to collect gases that are less dense than air.

- Downward delivery is used to collect gases that are denser than air.

- Gases that are fairly insoluble are collected over water.

- If we need to measure the volume of gas, we can collect it using a gas syringe.

Hazard Symbols

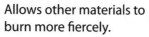
Oxidising

Allows other materials to burn more fiercely.

Highly flammable

Catches fire easily.

Toxic

Can cause death.

Harmful

Less dangerous than toxic.

Corrosive

Irritant

Attacks and destroys living tissue.

Can cause reddening or blistering of skin.

Hazard symbols warn us about chemicals:

Flame Tests

Flame tests can be used to identify the metals in compounds.

The colour of the flame indicates the **metal** present.

- lithium – red
- sodium – orange
- potassium – lilac
- calcium – brick red
- barium – apple green

Hydroxide Tests

We can identify the metals present in metal salt solutions by adding **sodium hydroxide solution**.

If the metal ion forms a precipitate, we can use the colour of the **precipitate** to identify the metal present.

- copper (II) – pale blue precipitate
- iron (II) – green precipitate
- iron (III) – brown precipitate

PROGRESS CHECK

1. Which gas burns with a squeaky pop?

2. What type of chemical allows other chemicals to burn more easily?

3. What type of chemical can cause death?

4. What does a corrosive chemical do?

5. What colour would you expect when a sample of a sodium salt is used in a flame test?

EXAM QUESTION

Use the words below to complete the sentences.

carbon dioxide   ammonia   chlorine   oxygen

If __a.__ is bubbled through limewater, the limewater turns cloudy. __b.__ relights a glowing splint. __c.__ bleaches litmus paper. __d.__ turns damp red litmus paper blue.

Rates of Reaction

The rate of reaction tells us how fast a chemical reaction takes place.

A chemical reaction takes place when:

- reacting particles **collide**

- colliding particles have enough **energy** to react – this is called the activation energy.

Temperature

When the **temperature** is increased:

- particles move faster

- they collide more often

- colliding particles have more energy, so more have enough energy to react

- the rate of reaction increases.

Concentration

When the **concentration** of a solution is increased:

- the particles collide more often

- so the rate of reaction increases.

Pressure

When the **pressure** of gases is increased:

- the particles collide more often

- so the rate of reaction increases.

Surface Area

When the **surface area** of a solid is increased (small pieces have a big surface area):

- the particles collide more often

- so the rate of reaction increases.

Catalyst

Adding a **catalyst** increases the rate of a chemical reaction. Catalysts are specific to certain reactions. Catalysts are not used up during reactions so they can be reused many times.

Following a Reaction

When magnesium metal is reacted with dilute hydrochloric acid, a salt called magnesium chloride and the gas hydrogen are made:

We can follow how fast this reaction happens by measuring:

- how quickly the hydrogen gas is made

- how quickly the mass of the reaction flask goes down as the hydrogen gas is made and escapes from the flask.

$$\text{magnesium} + \text{hydrochloric acid} \rightarrow \text{magnesium chloride} + \text{hydrogen}$$

This graph shows the amount of hydrogen produced in two experiments.

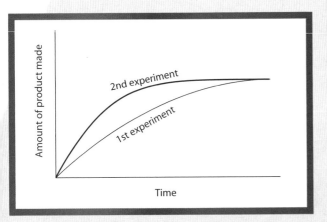

Time

1. What needs to happen for two particles to react?

2. How does increasing temperature affect the rate of a chemical reaction?

3. How does decreasing the concentration of a solution affect the rate of a chemical reaction?

4. How does adding a catalyst affect the rate of reaction?

5. During an experiment, when is the rate of reaction fastest?

■ In both experiments the rate of reaction is fastest at the start of the reaction.

■ The reaction is over when the line levels out.

■ The second experiment has a faster rate of reaction than the first experiment.

EXAM QUESTION

Use the words below to complete the sentences:

increases   catalysts   decreases   reactants

___a.___ increase the rate of a reaction but they are not used up themselves so they can be used many times. In the reaction between magnesium and hydrochloric acid increasing the surface area of the magnesium ___b.___ the rate of reaction. Decreasing the concentration of the hydrochloric acid ___c.___ the rate of reaction.

The reaction is over when one of the ___d.___ is used up.

Energy

In **exothermic** reactions, energy (normally in the form of heat) is transferred to the surroundings.

Most reactions are exothermic.

The fuel used in Bunsen burners is called methane. When methane is burnt it reacts with oxygen to release heat energy.

Neutralisation reactions are also exothermic. If sodium hydroxide solution is reacted with hydrochloric acid heat energy is released.

This means that if we recorded the temperature change during the reaction, we would see a temperature increase.

In **endothermic** reactions, energy (normally in the form of heat) is taken in from the surroundings. This means that if we recorded the temperature change during the reaction, we would see a temperature decrease.

Bonds

- Energy must be supplied to break **bonds**. Energy is released when new bonds are made.

- In exothermic reactions, more energy is released when new bonds are formed than is taken in to break old bonds.

- In endothermic reactions, more energy is taken in to break the old bonds than is released when the new bonds are formed.

The table below shows some average bond energies.

bond	Bond energy kJ mol^{-1}
C—C	347
O=O	498
C—H	413
O—H	464
C—O	358
H—Cl	432

Calorimetry

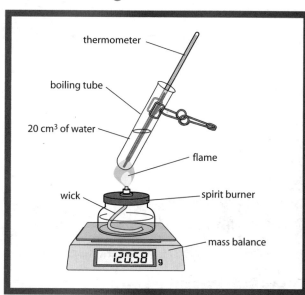

We can measure the energy released when fuels are burned using a technique called **calorimetry**.

In this process:

- We measure the volume of water in the boiling tube.

- We measure the temperature of the water in the boiling tube.
- We record the mass of the spirit burner and fuel.
- We light the spirit burner and the heat energy released as the fuel burns warms up the water in the boiling tube.
- When 1 gram of fuel has been burned, we turn out the spirit burner.
- The new temperature of the water is measured.

Calculating the Energy Transferred

We can calculate the heat energy transferred using the equation:

Heat energy transferred (J)
$$= \text{mass of water (g)} \times \text{specific heat capacity of water (J g}^{-1}\,^{\circ}\text{C}^{-1}) \times \text{change in temperature (}^{\circ}\text{C)}$$

The **specific heat capacity** of water is $4.2\ \text{J g}^{-1}\,^{\circ}\text{C}^{-1}$.

This means that it takes 4.2 J of energy to raise the temperature of 1 gram of water by 1°C.

In a calorimetry experiment 1g of fuel raised the temperature of 10g of water by 5°C.

The heat energy transferred is

Heat energy transferred (J) = $10\ \text{g} \times 4.2\text{J g}^{-1}\,^{\circ}\text{C}^{-1} \times 5^{\circ}\text{C}$
$= 210\ \text{J}$

To compare the amount of energy released by different fuels, we can divide the heat energy transferred by the mass of fuel burned.

PROGRESS CHECK

1. Is the burning of a fuel an exothermic or endothermic reaction?

2. If there is a temperature decrease during a chemical reaction, what type of reaction has taken place?

3. What is released when a new bond is made?

4. What is the name of the technique used to measure the amount of energy released when fuels are burned?

5. What is the unit used to measure the amount of heat energy transferred?

EXAM QUESTION

In a calorimetry experiment, 1 gram of fuel raised the temperature of 5 grams of water by 4°C.

a. Is the burning of this fuel an exothermic or an endothermic reaction?

b. Calculate the heat energy transferred during this reaction.

Producing Current

Electric **current** is a measure of how much **charge** flows in a certain amount of **time**.

Current is measured in **Amps** or Amperes. The number of Amps is equal to the amount of charge that flows every second.

The charged particles that flow in a wire are electrons which have a **negative charge**. They flow from negative to positive. Ions are positive particles left behind when the electrons move.

We say that electric current is the flow of **positively** charged ions from positive to negative, but ions do not actually move.

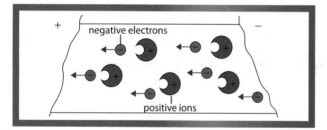

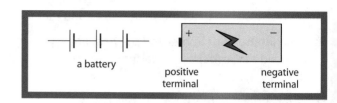

a battery

positive terminal

negative terminal

$$\text{Hours} = \frac{\text{Amp-hours}}{\text{Current supplied}}$$

For example, a battery is labelled 12 Amp-hours and it is used to supply 0.5 A to a CD player.

$$\frac{\text{12 Amp-hours}}{\text{0.5 A}} = \text{24 hours}$$

The battery will last for 24 hours. Rechargeable batteries produce less waste which is better for the environment.

Cells

Cells produce a flow of charge when a wire is connected across the terminals. The current flows from the positive terminal of the cell to the negative terminal. The flow of current from a cell is always in one direction, it is called **direct current** or **DC**. Solar cells also produce direct current.

A **battery** is a number of cells connected together in series. A battery's capacity is stated in Amp-hours. The number of hours a battery will last depends on its capacity in Amp-hours and the current it is supplying. Different appliances will take different currents from the same battery.

Generators

Electricity is generated in a power station by rotating a magnet near a coil of wire (or by rotating a coil of wire near a magnet). This produces a current in the coil of wire and is called the **dynamo effect**.

Dynamos or generators produce a current that changes direction every time the magnet turns. This is called **alternating current** or **AC**. The current supplied to our homes is AC.

The current and the voltage of a dynamo can be increased by:

■ increasing the strength of the magnet

■ increasing the number of turns in the coil of wire

■ increasing the speed of rotation of the magnet.

AC and DC Current

AC and DC current can be displayed on an oscilloscope.

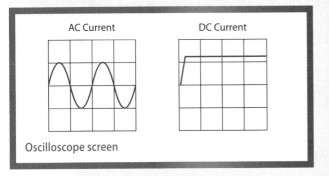

AC Current | DC Current

Oscilloscope screen

Measuring Current

The current in a circuit is measured using an **ammeter**. The ammeter must be placed in the circuit in **series** (next to) any other components. The current is the same at all points in a series circuit so it can be placed either side of the component.

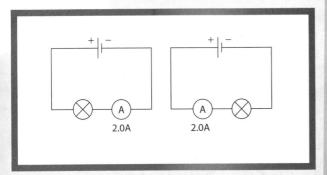

PROGRESS CHECK

1. How is current defined?

2. Are the following statements true or false?

   a. Electrons have a negative charge.

   b. Electrons flow from positive to negative.

   c. Ions flow from positive to negative.

   d. Current is said to flow from positive to negative.

EXAM QUESTION

The diagram shows a simple generator. When the coil is rotated, a current flows and lights the lamp.

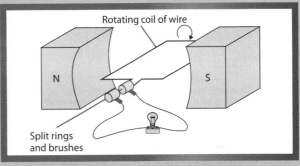

a. Give **three** ways of increasing the current in the coil.

b. Is the current produced AC or DC?

c. Explain the difference between AC and DC.

d. Add to the diagram an ammeter that would measure the current passing through the lamp.

Resistance

Current passes easily through copper wire. Copper wire has a low **resistance**.

Current does not pass so easily through a filament lamp. A filament lamp has a higher **resistance**. More energy is needed to push the electrons through the filament wire in the lamp. This energy is converted to heat (and light) in the lamp. Components with a higher resistance give off more heat.

Calculating Resistance

Resistance is measured in Ohms (Ω). The resistance of a component is found using the following equation:

$$\text{Resistance} = \frac{\text{Potential difference}}{\text{Current}}$$

- Resistance in Ohms (Ω)

- Potential difference (voltage) in Volts (V)

- Current in Amps (A)

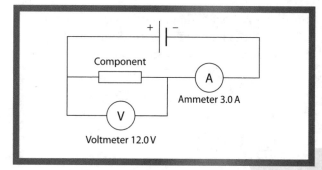

There is a potential difference across the component above of 12.0 V and a current through it of 3.0 A.

$$\text{Resistance} = \frac{12.0\,\text{V}}{3.0\,\text{A}} = 4.0\,\Omega$$

- If the potential difference remained the same, a component with a higher resistance would have a **lower** current passing through it.

- If the potential difference remained the same, a component with a lower resistance would have a **higher** current passing through it.

Changing Resistance

The resistance of some components is constant. The resistance of other components can change.

- The resistance of a filament lamp increases as more current passes through it. As the current increases, the lamp gets hotter.

or

- A light dependent resistor (LDR) changes resistance according to the amount of light. An LDR has a **high** resistance in the dark and a **low** resistance in the light. An LDR can be used to detect light and switch a lamp on automatically in the dark, or to control the shutter speed in a camera.

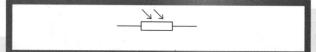

- A thermistor changes resistance according to the temperature. A thermistor has a **high** resistance when cold and a **low** resistance when hot. A thermistor can be used to detect temperature change for a fire alarm.

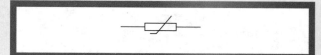

Graphs of Potential Difference Against Current

If potential difference for a component is plotted on the y-axis and current on the x-axis, the gradient of the graph is equal to the resistance.

Graph for a Component with a Constant Resistance

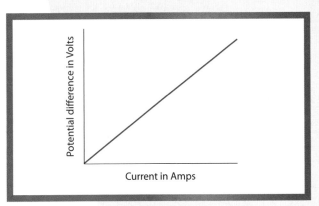

The gradient is constant.

Graph for a Filament Lamp

The gradient increases as the current increases.

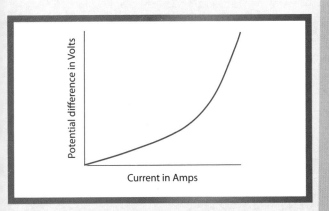

PROGRESS CHECK

1. Find the resistance of a component with a p.d. of 6.0 V and a current of 4.0 A.

2. A resistor is replaced with a resistor of higher resistance. How will the current in the circuit change?

3. What is the circuit symbol for a light dependent resistor?

4. Explain how the resistance of a filament lamp changes as the current through it increases.

5. What does the V-I graph look like for a component with constant resistance?

? EXAM QUESTION

A thermistor is a device used to detect temperature change.

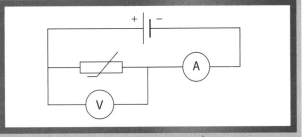

a. In the above circuit, the measurements of the voltmeter and the ammeter are 15.0 V and 2.5 A respectively. Calculate the resistance of the thermistor.

b. The thermistor is placed in a beaker of crushed ice. What happens to the resistance of the thermistor?

c. Assuming that the p.d. in the circuit remains the same, how would the current change when the thermistor is in the crushed ice?

d. Suggest a use for a thermistor.

Power

Electrical **power** is a measure of the rate at which a device uses **energy**. An appliance that uses more energy per second has more power.

Power is measured in **watts** (W). An appliance that uses 1 joule of energy in 1 second has a power of 1 watt.

A light bulb with a power of 60W uses 60J of energy per second. A light bulb with a power of 100W uses 100J of energy per second.

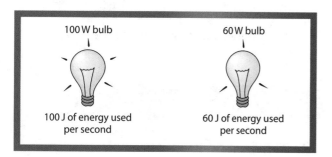

100W bulb   60W bulb

100J of energy used per second   60J of energy used per second

$$\text{Power} = \frac{\text{Energy}}{\text{Time}}$$

- Power in watts (W)
- Energy in joules (J)
- Time in seconds (s)

Energy
—————
Power × Time

Calculating Electrical Power

Power is calculated using the following equation:

Power = potential difference (voltage) × current

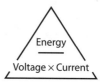

Energy
—————
Voltage × Current

Example

An electric drill operates on a voltage of 230V and has a power of 400W.

Calculate the current drawn by the drill.

$$\text{current} = \frac{\text{power}}{\text{voltage}}$$
$$= \frac{400\,W}{230\,V}$$
$$= 1.7\,A$$

Example

A hairdryer of 1200W (1.2kW) is used for 2 minutes.

How much energy is used?

$$
\begin{aligned}
\text{energy} &= \text{power} \times \text{time} \\
&= 1200\,W \times 120\,s \\
&= 144\,000\,J = 144\,kJ
\end{aligned}
$$

Domestic Bills

■ The standard unit for energy is joules.

■ A typical appliance uses a large number of joules, the hairdryer in the example on page 86 uses 144 000 J in just 2 minutes!

■ For domestic bills, it is more convenient to use a larger unit. The **kilowatt-hour** is a larger unit of **energy**.

■ An appliance with a power of 1000 W (1 kW) switched on for 1 hour uses 1 kilowatt-hour of energy. (Compare this to an appliance with a power of 1 W switched on for 1 s which uses 1 J of energy.)

Example

A fridge of 1.5 kW is switched on for 8 hours. How many kWhs does it use?

energy = power × time

$\qquad\qquad$ = 1.5 kW × 8 hours

$\qquad\qquad$ = 12 kWhs

Electricity companies refer to kWhs as **units** of energy.

The cost can be calculated by multiplying the number of units by the cost of a unit.

cost = number of units (kWhs) × cost of a unit

◉ PROGRESS CHECK

1. A personal stereo produces 1.5 J of sound energy per second. What is its power?

2. Calculate the energy (in Joules) used by a 2000 W kettle in 2 minutes.

3. Calculate the energy in kWhs for the same kettle for the same time.

4. A lamp is marked with the following information: 12 W, 2.5 V.
   Calculate the current through the lamp.

5. An electricity company charges 12 p for each unit of electricity used.
   Calculate the cost of operating a 2 kW heater for 4 hours.

? EXAM QUESTION

A table lamp has three bulbs, each of 40 W.

a. Find the energy in Joules and in kWhs used by each bulb per hour.

b. If each unit costs 10 p, find the cost of lighting the table lamp for 24 hours.

c. It is suggested that the bulbs are replaced with low energy bulbs of 5 W each, giving out the same light, but less heat. How much energy in KWhs is used by each bulb per hour?

d. Calculate the kWhs used by the lamp in 24 hours with low energy bulbs and find the cost of lighting the table lamp for 24 hours.

Using Electricity

Electricity is used in many everyday devices.

Some examples of everyday devices designed to bring about particular energy transformations are listed below.

- A hairdryer converts electrical energy to kinetic energy and heat.
- A microphone converts sound (kinetic energy of the air) to electricity.
- A light bulb converts electricity to light and heat.

The development of the use of electricity and telecommunications has had a great impact on society. Here are some of the ways that things have changed over time.

- The increased use of electricity and electrical devices has had a dramatic effect on the way we live.
- Electric circuits have become much smaller, greatly increasing the processing speed of computers and allowing development of many new applications.
- The electric telephone has replaced rotary dialling with touchtone dialling and greatly increased the capacity and quality of the network. Digital signalling has replaced analogue, allowing multiplexing (sending many signals at once).
- The use of ICT in collecting and displaying data for analysis has improved reliability and validity of data and encourages faster development of new technology.
- Maglev trains suspend, propel and direct trains using magnetic forces and magnetic levitation from electromagnets attached to the train. The decreased friction allows trains to reach speeds of nearly 600 km/h.

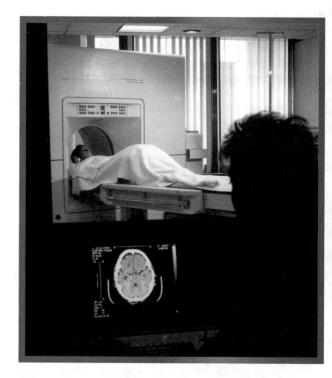

- Superconductivity is the decreasing of resistance to almost zero in certain materials at extremely low temperatures. The technology is used to make very powerful electromagnets used in MRI machines and particle accelerators.

Safe Electricity

Some ways of improving safety when using electricity are listed below.

- A **fuse** is a short piece of thin wire placed in a circuit that melts and breaks the circuit if the current is too high.

- **Residual current circuit breakers** (RCCBs) improve the safety of a device by disconnecting a circuit automatically whenever it detects that the flow of current is too high.

- An **earth wire** is a safety wire that connects the metal parts of a device to earth and stops the device becoming dangerous to touch if there is a loose wire. If a wire works loose and touches a metal part of the device, a large current flows to earth, blowing the fuse or activating the circuit breaker.

- Many modern devices are insulated with a plastic case which insulates the user from any wiring faults. Since the wires themselves are also insulated, this is known as **double insulation**.

PROGRESS CHECK

1. How has the size of electric circuits affected computers?

2. What is multiplexing?

3. How can a maglev train reach speeds of almost 600 km/h?

4. What is superconductivity?

5. What is a fuse?

EXAM QUESTION

This question is about an electric hairdryer.

a. Name the energy change that takes place in a hairdryer.

b. What is an RCCB?

c. Explain how someone using a hairdryer is protected by the use of an earth wire and a fuse.

d. A hairdryer that is double insulated does not require an earth wire. Explain why.

Harnessing Electricity

Electricity is supplied to our homes at 230V, but some devices, such as printers and mobile phone chargers, require less voltage.

Transformers

Transformers change the voltage of an **alternating** electricity supply.

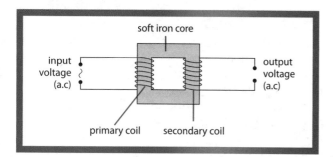

- A **primary** coil is an electric cable wrapped around a soft iron core. It is connected to an input power supply providing **alternating** current.

- The alternating current creates a **changing** magnetic field in the soft iron.

- The changing magnetic field in the soft iron core causes a voltage to be **induced** in the **secondary** or output coil.

- The number of turns of the primary and the secondary coil determines the change in voltage according to the following equation:

$$\frac{\text{primary or input voltage}}{\text{secondary or output voltage}}$$

$$=$$

$$\frac{\text{number of turns on primary coil}}{\text{number of turns on secondary coil}}$$

The National Grid

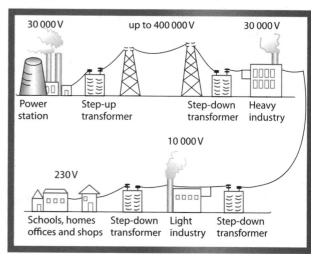

- The National Grid distributes electricity to consumers at different voltages using **step-up** and **step-down** transformers. Consumers include homes, factories, offices, schools and farms.

- It is more efficient to transmit electricity at a much higher voltage than is required by consumers. Increasing the voltage decreases the current in the transmission cables. The cables lose less heat energy when the current is lower.

- Overhead cables are the cheapest way of transmitting electricity. Underground cables are more expensive, but do not spoil the landscape.

- The National Grid allows electricity to be distributed from areas where consumption is low to areas where there is high consumption.

Motors

Motors take electrical energy and transform it into kinetic (moving) energy.

The simple motor below runs on DC (direct current) from a battery. The current flows in the coil, which experiences a turning effect caused by the magnetic field.

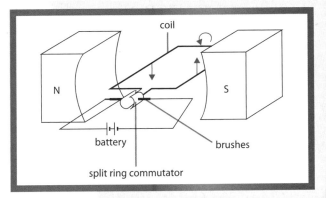

The forces on the coil can only pull the coil half a turn because the forces then pull the wrong way. A motor allows the coil to continue to turn by changing the direction of the forces. It does this by changing the direction of the current, as the coil spins, the commutator spins, but the brushes remain in the same place, changing the direction of the current every half turn.

A stronger turning effect is produced by:

- increasing the number of turns on the coil

- increasing the current

- using a stronger magnet

- using a bigger coil.

👁 PROGRESS CHECK

1. The input voltage of a transformer is 230 V and the primary coil has 1200 turns. If the secondary coil has 100 turns, what is the output voltage?

2. Why does the National Grid transmit electricity at high voltages?

3. Give an example of a 'high voltage'.

4. Give **one** advantage of using underground cables to distribute electricity.

5. Give **three** ways of increasing the turning effect of a motor.

❓ EXAM QUESTION

The output of a transformer is 20 V and the input is 200 V.

a. What is the ratio of primary turns to secondary turns?

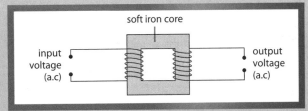

b. Label the primary coil and the secondary coil in the diagram above.

c. Is this a step-up transformer or a step-down transformer?

d. Add a voltmeter to the diagram that would measure the output voltage.

Harnessing Energy I

Most of our energy comes from fossil fuels such as oil or gas. Fossil fuels are non-renewable, so, eventually, they will run out.

Fossil fuels release gases into the environment that pollute the atmosphere.

The energy change when burning fossil fuels is from **chemical energy** to **heat** to **kinetic energy** (in a turbine which turns a generator) to **electrical energy**.

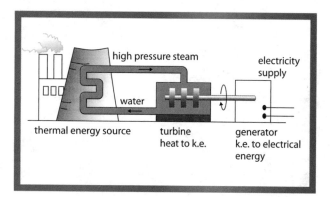

high pressure steam

electricity supply

thermal energy source

water

turbine
heat to k.e.

generator
k.e. to electrical
energy

Energy can be harnessed from sources such as the Sun, the force of the wind or moving water. Such energy sources are renewable, they require no fuel and release no polluting gases into the environment.

There are often difficulties associated with renewable energy sources, however, including other problems for the environment. Many renewable energy sources are also unreliable and expensive. Many people think we should aim to reduce the amount of energy we use in our homes and workplaces.

Wind Power

A wind farm is a collection of generators driven by wind turbines. The generators are used to produce energy.

Some of the problems associated with wind farms are as follows.

- They require large areas of land.
- They are dependent on wind speed.
- The noise can disturb wildlife.
- They can be expensive to build.
- They change the view of the landscape.

The energy change is from **kinetic energy** of the wind to **electrical energy**.

Water Power

Waves and Tides

Energy can be harnessed from the rise and fall of water due to waves and tides. The **kinetic energy** of the water can be used to drive turbines, which generate **electrical energy**.

A disadvantage is that wave and tidal barrages are very difficult to build.

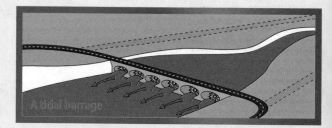

A tidal barrage

Hydroelectric Power

Rivers and rain fill up a high reservoir behind a dam. The water is released and used to drive turbines which generate electricity. The energy changes are **gravitational potential energy** (of the water) to **kinetic energy** (of the water) to **kinetic energy** (of the turbine) to **electrical energy**.

Harnessing energy from water can disturb natural habitats. Tidal barrages and dams flood areas and remove water from other areas artificially.

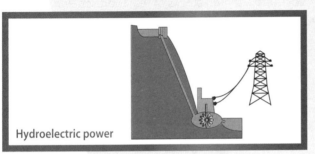

Hydroelectric power

Geothermal Energy

In some volcanic areas, hot water and steam rise to the surface. Water can also be pumped down to hot rocks and steam is generated. The steam can be used to drive turbines. The energy change is from **heat** to **kinetic energy** to **electrical energy**.

The disadvantage of this source of energy is that deep drilling for hot rocks is expensive and difficult.

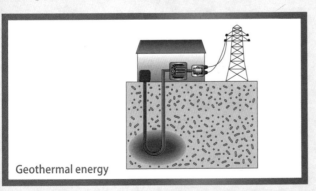

Geothermal energy

PROGRESS CHECK

1. Name the energy changes in a power station burning fossil fuels.

2. Give **three** renewable energy sources.

3. Name an advantage of hydroelectric power.

4. Name a disadvantage of hydroelectric power.

5. Name the energy change that takes place when using geothermal energy.

EXAM QUESTION

This question is about wind power.

a. Is wind power renewable or non-renewable?

b. Name **two** advantages of wind power.

c. Name **two** disadvantages of wind power.

d. What is the energy change in a wind turbine?

Harnessing Energy 2

Many alternative energy sources disrupt or damage the environment, disturbing wildlife.

Biomass

Fuels can be made from plant or animal matter and these are used to heat water. The steam is used to drive a turbine.

Biofuels require large areas of land to grow the plants.

Solar Power

Solar **cells** convert energy from the Sun into electricity. Solar **panels** use infrared radiation from the Sun to heat water.

Solar cells are very expensive and sunshine is not reliable in many locations.

Nuclear Power

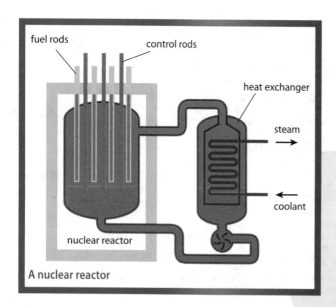

A nuclear reactor

Radioactive fuel rods, such as uranium, release energy as heat through a process called nuclear fission.

A nuclear power station uses this heat to drive turbines and generate electricity.

Some of the advantages and disadvantages of nuclear power are:

- Nuclear power is independent from fossil fuels and does not contribute to global warming.
- There are high stocks of fuel (non-renewable).
- Small amounts of fuel give large amounts of energy.
- It has high maintenance costs.
- It has high decommissioning (closing down) costs.
- There is a risk of accidental emission of radioactive material.
- Waste material disposal is expensive.

Low level radioactive waste can be buried in landfill sites. Other waste must be encased in glass or concrete before it is buried.

It is possible to reprocess some types of waste. Some problems of dealing with radioactive waste are listed below.

- It can cause cancer if it is not disposed of correctly.
- It remains radioactive for a long time.
- Plutonium (waste product from nuclear reactors) can be used to make bombs and could be considered a terrorist risk.
- It must be kept out of groundwater.
- Acceptable levels of radioactivity may change over time.

Environmental Issues

■ Power stations that burn fuel put carbon dioxide gas into the atmosphere. The gases act like a greenhouse and trap energy from the Sun which may cause global warming. This greenhouse effect is important – without it our world would not be warm enough for life to exist. Extra warming, however, could cause problems for humans, plants and animals. Deforestation and increased CO_2 also contribute to global warming.

■ Some power stations give off other gases such as sulfur dioxide, which can cause acid rain. Acid rain damages buildings and stone.

■ A radioactive leak could cause damage to humans and wildlife for many years. Radioactive dust can be carried by the wind for thousands of kilometres.

■ The ozone layer protects the earth from ultraviolet radiation and pollution from CFCs is depleting the ozone layer.

■ Dust from volcanoes reflects radiation from the Sun which causes cooling, dust from factories reflects radiation from the city which causes warming.

PROGRESS CHECK

1. State **one** disadvantage of biomass fuels.

2. What chemical causes acid rain?

3. What gas is said to cause the greenhouse effect and global warming?

4. Which gases damage the ozone layer?

5. How does dust from factories affect the climate?

EXAM QUESTION

A nuclear power station provides energy for a city.

a. Name a fuel used in nuclear power stations.

b. Give **two** advantages of nuclear power.

c. Give **two** disadvantages of nuclear power.

d. Name **two** risks associated with nuclear power.

Keeping Warm

Heat travels from hot places to cold places. This generally results in hot things cooling down (losing their heat) and cold things heating up (gaining heat).

A warm house in winter can be kept warm using insulation.

Heat can move in three ways.

1. **Conduction** involves the heat moving from one particle to another 'usually in a solid.' Some materials are better thermal conductors than others. Metal is a good conductor since it allows the heat to travel through it quickly. Plastic and wood are poor conductors, or insulators, so are most liquids. Many good thermal insulators have pockets of air trapped in them.

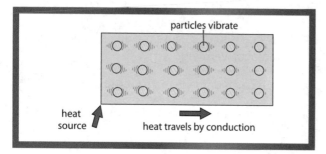

particles vibrate

heat source

heat travels by conduction

2. **Convection** involves the movement of particles in a gas or a liquid. Thermals are areas of hot air rising, cooler air sinks. The air circulates in a **convection current**. Heat circulates from a heater around a room by convection currents. Convection cannot take place in a solid since the particles are not free to move. Heat transfer by convection can be reduced by trapping air such as in double glazing.

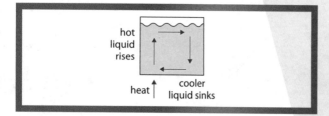

hot liquid rises

heat

cooler liquid sinks

Hot air rises because it expands and becomes less dense (has less mass per unit volume). Cool air sinks because it is more dense.

3. **Radiation** is the transfer of heat energy by **infrared electromagnetic waves**. All bodies emit and absorb thermal radiation. The hotter a body is, the more energy it radiates. Dark, matt objects absorb more radiation than light, shiny ones. This results in dark objects heating up more and emitting more radiation, too. Radiation does not require particles so it can travel through a vacuum. This is how the Sun's energy reaches Earth.

Here are some ways that a house can be insulated. You will notice that many of them trap air and therefore reduce heat transfer by convection.

- Double glazing traps air between two layers of glass.

- Curtains trap air.

- Loft insulation reduces hot air rising and escaping through the roof.

- Cavity walls trap air – cavities in walls are often filled with insulation which further reduces the movement of the air.

- Draught proofing.

- Reflective foil on or in walls reduces heat loss by radiation.

- Insulation (lagging) around the hot water tank.

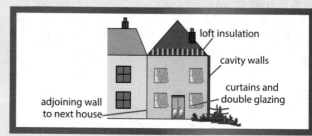

loft insulation

cavity walls

curtains and double glazing

adjoining wall to next house

Note that a well insulated house will also keep cooler during a hot summer.

Here are some other factors to take into account when considering the effectiveness of saving energy in a home.

- The thickness of the walls.

- The area and number of windows.

- A detached house has more surface area than a terraced house.

PROGRESS CHECK

1. Name the **three** ways that heat can move.

2. Which type of heat transfer occurs best in solids?

3. Why does hot air rise?

4. What sort of body absorbs and emits more radiation?

5. Give **three** ways of improving the insulation of a house.

? EXAM QUESTION

The diagram shows a cavity wall which is used to improve the insulation of a house.

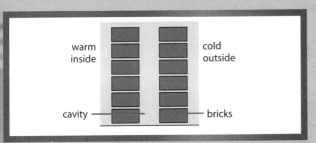

a. By what method does heat flow through the bricks?

b. Draw arrows to show possible convection currents in the cavity.

c. Suggest how the insulation of this system could be improved.

d. Does a cavity wall keep the inside of a house warm or cool when the weather is hot?

Efficiency

All devices waste energy; this energy is usually lost as heat to the surroundings. Efficiency is a measure of how much energy is lost and how much is used.

Your body uses energy. If you run up a flight of stairs, you may use 2000 Joules of energy. This energy comes from stored energy in the food you eat.

Only about 15% of this energy, however, is transferred to the movement in your muscles, the rest is wasted as heat in your body and heat to the surroundings. So about 300 J would be transferred to kinetic energy and 1700 J to heat.

$$\text{Efficiency} = \frac{\text{useful energy output}}{\text{total energy input}} \times 100\%$$

Total energy input   = 2000 J
Useful energy output   = 300 J

$$\text{Efficiency} = \frac{300}{2000} \times 100\% = 15\%$$

The sankey diagram/flow chart below shows the energy efficiency for a person running up a flight of stairs.

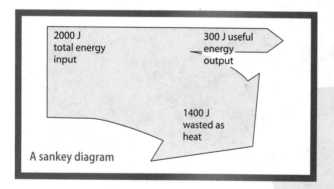

2000 J total energy input

300 J useful energy output

1400 J wasted as heat

A sankey diagram

Here are some typical efficiencies of other devices:

- An electric motor   80%
- A petrol engine   25%
- A candle   5%
- A car engine   25%

As a petrol engine loses 75% of its energy as heat, a car needs a cooling system such as a radiator.

A hairdryer might be 75% efficient because the heat from a hairdryer is useful. The rest of the energy is converted into sound and kinetic energy.

A sankey diagram/flow chart for a hairdryer shows how the electrical energy is converted into heat, sound and kinetic energy.

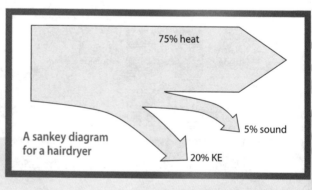

75% heat

5% sound

20% KE

A sankey diagram for a hairdryer

Power can also be used to calculate efficiency. For example, a conventional light bulb may use 100 W of power. This means that it converts 100 J of electrical energy to light and heat energy every second. A typical amount of heat given off by a conventional light bulb is 90 W of power. This means that the light bulb converts 90 J of electrical energy to heat energy per second.

The efficiency of the light bulb can be calculated using power as follows.

$$\text{Efficiency} = \frac{\text{useful energy output}}{\text{total energy input}} \times 100\%$$

$$= \frac{10\,\text{W}}{100\,\text{W}} \times 100\%$$

$$= 10\%$$

This works because power is a measurement of energy changed per second, so for any specific amount of time, the useful energy input and the total energy output are proportional to the power.

As energy is used, it tends to become more spread out. When it is more spread out, it is more difficult to use. Eventually, all energy, both wasted and useful, is transferred to its surroundings, which become warmer.

PROGRESS CHECK

1. State the equation for efficiency.

2. What is the efficiency of a machine that uses 40 J and gives out 20 J of work?

3. What is a typical efficiency of a human body?

4. Why do cars need cooling systems?

5. What is the efficiency of a device that has an input power of 200 W and an output power of 70 W?

? EXAM QUESTION

The diagram on the right shows the flow of energy through a 60 W light bulb.

a. Explain why the light bulb uses 60 J of energy per second.

b. What kind of energy does the light bulb use?

c. If the light bulb gives out 5 J of light per second, calculate its efficiency.

d. What happens to the rest of the energy?

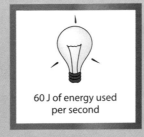

60 J of energy used per second

Heat

Nearly all of our energy comes from the Sun. Solar panels use energy from the Sun to heat up water in pipes to use for heating.

Trapping Heat

Solar panels are painted black to increase the absorption of radiation.

Solar cells convert energy from the Sun directly to electricity. **Photocells** convert the Sun's energy directly into electricity.

Some information about photocells is given below.

- Photocells work by absorbing energy which knock electrons loose from silicon atoms in the crystals.

- Electrons then flow freely to produce direct current.

- The power of photocells depends on their surface area and the light intensity.

- Photocells can operate in remote locations.

- Photocells produce no waste.

- Photocells are low maintenance.

- Photocells do not work in bad weather or at night.

Passive solar heating works because glass allows the Sun's radiation to pass through but it reflects the infrared radiation given off by the heated surfaces inside.

Temperature and Heat Transfer

Temperature is a measurement of hotness and is measured in degrees Celsius (°C). Heat is a measurement of energy and is measured in Joules (J). Temperature can be represented by a range of colours in a thermogram (see photograph below).

Factors Affecting Heat Transfer

Here are some factors that affect the rate at which heat is transferred.

- If a body is warmer (at a higher temperature) than its surroundings, it will lose heat energy to its surroundings. The greater the temperature difference, the faster the rate at which heat is transferred.

- If a body is cooler (at a lower temperature) than its surroundings, it will gain heat energy from the surroundings. The greater the temperature difference, the faster the rate at which heat is transferred.

- The shape and the dimensions of a body affect the rate at which it transfers heat. For example, a body with a larger surface area will lose heat faster. Babies have a large surface area compared with their mass; this is why it is more difficult for babies to keep warm.

- Under similar conditions, different materials transfer heat at different rates. In a cold room, a carpet does not transfer heat very quickly (it is a good thermal insulator). If you walk on the carpet without shoes you will not feel cold because the heat is being transferred away from your warmer feet at a slow rate. However, if you walk on a tiled floor at the same temperature, your feet will feel cold. This is because the tiles transfer heat away from your feet faster (they are good thermal conductors).

Energy is needed to make a solid melt and to make a liquid evaporate. This energy is absorbed from the environment.

When a substance condenses or freezes, it loses energy to its surroundings.

PROGRESS CHECK

1. Why are solar panels painted black?

2. What is temperature?

3. What is heat?

4. Why is it difficult for babies to keep warm?

5. Why does a tiled floor feel cold?

? EXAM QUESTION

This question is about solar cells.

a. What happens to the electrons in a solar cell when it absorbs energy?

b. What factors affect the amount of energy absorbed?

c. Name **one** advantage of solar cells.

d. Name **one** disadvantage of solar cells.

Heat Calculations

The energy needed to change the temperature of a body depends on its mass, the material it is made from, the temperature change required.

Specific Heat Capacity

The **specific heat capacity** tells us how much energy 1 kg of a material needs to increase its temperature by 1 °C. The unit of specific heat capacity is Joules per kg per degree.

The energy required to heat a substance, or the heat lost by a substance when cooling, is calculated using the following equation:

$$\text{Energy} = \text{mass} \times \frac{\text{specific heat capacity}}{} \times \frac{\text{temperature change}}{}$$

Example

500 g of water is allowed to cool from 80 °C to 55 °C.

How much energy is lost to its surroundings (specific heat capacity of water is 4200 J/kg/°C)?

$$\text{Energy} = 0.5 \text{ kg} \times 4200 \text{ J/kg/°C} \times 25\,°C$$
$$= 52\,500 \text{ J}$$

When heating a substance, some of the energy is lost to the surroundings. Therefore we calculate the **minimum** amount of energy required.

Specific Latent Heat

If we heat ice at −20 °C at a constant rate, its temperature increases at a steady rate until it reaches 0 °C. Then the ice continues to absorb energy at the same rate without getting any hotter. It is using the energy to break down the bonds between the ice molecules as it melts.

The energy required to melt a substance is the **latent heat**. Liquid requires latent heat, or heat energy to evaporate.

Gases lose heat as they condense and liquids lose heat as they freeze. Energy is required to change state; all changes happen at constant temperature.

The **specific latent heat** of a substance is the amount of energy required to change the state of 1 kg of that substance at constant temperature.

We can calculate latent heat using the following equation:

$$\text{Energy} = \text{mass} \times \text{specific latent heat}$$

Graphs

As a substance is heated or cooled, a graph can show the change in temperature. The graph below shows ice as it is heated and melts and then heated (as water) then evaporates. The heating takes place at a constant rate, so rate of heat added is proportional to the time.

The horizontal section from 7 to 10 minutes shows that the ice is absorbing heat energy as it changes from ice to water at a constant temperature of 0 °C. Between 10 minutes and 13 minutes the substance is water. The horizontal section from 13 minutes to 16 minutes shows the water absorbing heat as it evaporates at 100 °C.

PROGRESS CHECK

1. What is specific heat capacity?

2. At what temperatures does water freeze and melt?

3. What is specific latent heat?

4. Calculate the minimum energy needed to raise the temperature of 2 kg of water by 5 °C.

5. Calculate the energy lost when 1 kg of copper cools by 5 °C (shc = 385 J/kg/°C).

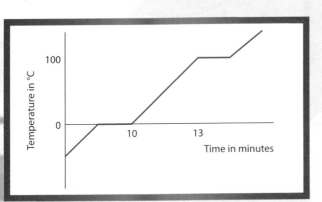

? EXAM QUESTION

Look at the graph above.

a. Explain the horizontal line at 0 °C.

b. Calculate the minimum energy required to heat the water from 0 °C to 100 °C (mass = 1 kg).

c. Why is this the minimum energy required?

d. Calculate the energy required to vaporise the water (latent heat of vaporisation = 2 260 000 J/kg).

Radiation

The basic structure of an atom is a small central **nucleus** composed of **protons** and **neutrons** surrounded by **electrons**.

- Protons have a positive charge, electrons have a negative charge and neutrons have no charge: they are neutral.

- The number of protons a nucleus has tells us what kind of element the atom is. Atoms of the same element always have the same number of protons.

- An atom with an overall neutral charge will have the same number of electrons as it has protons.

- An atom with extra electrons has an overall negative charge. An atom with fewer electrons has an overall positive charge.

- An atom with a different number of neutrons will have a different mass, although it is still the same element. Atoms with different numbers of neutrons are called isotopes. An isotope of an element has the same number of protons but extra or fewer neutrons.

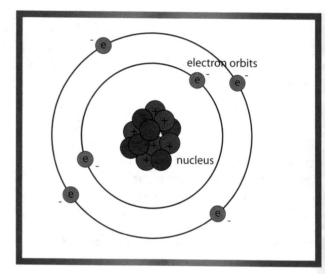

When a nucleus is unstable it may emit some of its particles. Atoms that give out radiation from their nucleus all the time are said to be **radioactive**.

- **Alpha** radiation is a helium nucleus consisting of two protons and two neutrons. It is emitted as a particle straight from the nucleus.

- **Beta** radiation is an electron that is emitted from the nucleus. No electrons exist as electrons in the nucleus. However, if a nucleus is unstable a neutron may spontaneously change into a proton inside the nucleus. When it does this, it emits an electron. This is called beta radiation.

- **Gamma** radiation is the emission of an electromagnetic wave from the nucleus of an atom. It usually follows alpha or beta radiation.

Properties of Radiation

Some properties of these types of radiation are in the table below.

	Alpha	Beta	Gamma
Range in air	A few centimetres	About a metre	Huge distances
Penetration power	Stopped by a thick sheet of paper or skin	Stopped by a few centimetres of aluminium or other metal	Lead or thick concrete will reduce its intensity
Ionising power	Strong	Weak	Very weak
Charge	+2	−1	None
Deflection in an electric or a magnetic field	Deflected but not a lot due to their heavy mass	Deflected a lot due to their low mass	No deflection

Ionisation

Ionisation is the ability of the radiation to cause other particles to gain or lose electrons. Ionisation produces charged particles. Radiation can be detected using a Geiger Muller tube which uses ionisation to detect the radiation.

PROGRESS CHECK

1. What is the charge on a neutron?

2. Explain how a beta particle can be an electron emitted from a nucleus.

3. What is the range of gamma radiation in air?

4. Are beta particles deflected in a magnetic field?

5. What is ionisation?

EXAM QUESTION

An isotope of uranium is called uranium 235. The number 235 indicates the number of neutrons in the nucleus of each atom.

a. What is the name given to atoms of the same element with different numbers of neutrons?

b. A uranium nucleus decays with alpha decay. What is the nature of an alpha particle?

c. A detector is held 2.0 m away from an alpha source. Explain why the detector will not detect the radiation.

d. Suggest a suitable test to determine whether a source emits alpha particles.

Radiation in our World

Background radiation is a low level of radiation that is always around us.

Background Radiation and Half-life

Background radiation is mainly caused by natural radioactive substances such as rocks, soil, living things and cosmic rays. Humans also contribute to background radiation, for example, through medical uses, nuclear waste and power stations.

The **half-life** of a radioactive isotope is defined as the time it takes for the number of nuclei in a sample to halve.

The **half-life** of a radioactive sample can also be defined as the time it takes for the count rate of a sample (number of radiations per second) to decrease by half.

Risks

Radiation can damage or destroy living cells due to its ionising power. It can cause cancer or make vital organs stop working.

- Alpha radiation is highly ionising and very dangerous if it is taken into the body in food or by breathing in radioactive gas or dust. Once absorbed, the radiation can cause damage deep inside the body.

- Beta and gamma rays can penetrate the skin, but most sources of these radiations are well shielded, such as power stations and laboratories.

- Some media reports claim that microwave radiation from mobile phones or masts poses a health risk.

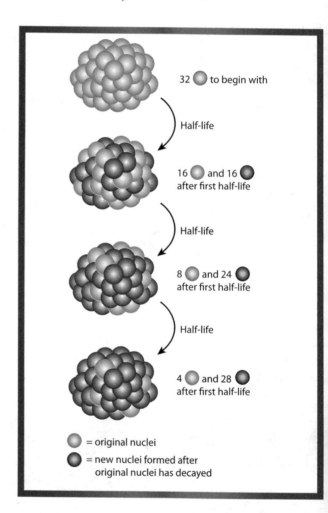

32 ⚪ to begin with

Half-life

16 ⚪ and 16 ⚫ after first half-life

Half-life

8 ⚪ and 24 ⚫ after first half-life

Half-life

4 ⚪ and 28 ⚫ after first half-life

⚪ = original nuclei
⚫ = new nuclei formed after original nuclei has decayed

Safety in the Laboratory

Radioactive substances must be handled safely.

- Use tongs or gloves.
- Minimise the exposure time.
- Store sources in shielded containers.
- Wear protective clothing.

Uses of Radiation

Radiation is used in a number of ways.

- Alpha radiation is used in smoke detectors.

- Beta radiation can be used to detect the thickness of paper or metal sheet.

- Beta radiation can be used as a tracer to detect leaks in underground pipes.

- Radioisotopes can be used in medicine as tracers. They can be added to a liquid and tracked in the body. Radioisotopes are chosen that have low penetration power and a low half-life (within the realms of practicality) to reduce the risk.

- Gamma rays are used to treat cancer.

- Gamma rays are used to sterilise equipment.

 **PROGRESS CHECK**

1. State **three** sources of background radiation.

2. Give a definition of half-life.

3. Give a reason why alpha radioactive dust is very dangerous.

4. Suggest a use for alpha radiation.

5. Suggest one use for gamma radiation.

? **EXAM QUESTION**

The diagram to the right shows a radiation source being used to regulate the thickness of a metal sheet.

a. Label the detector in the diagram.

b. Suggest whether the source is alpha, beta or gamma.

c. Suggest another use for this type of radiation.

d. When handling a radioactive source name **two** precautions that should be taken.

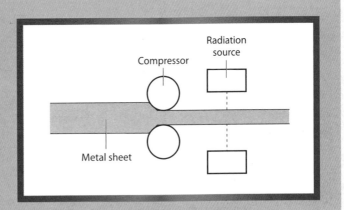

Waves

Waves transmit energy from one place to another either through space or through a material.

Types of Waves

The material itself is not transmitted by the wave, only displaced and then it returns to its original position.

- **Frequency** is the number of waves passing a point per second and is measured in Hertz (Hz).

- **Wavelength** is the distance between any point on a wave and the same point on the next wave.

- **Amplitude** is the maximum displacement of the waves from rest position.

There are two types of wave: **longitudinal** and **transverse**.

The wave equation states that for any wave:

$$\text{speed} = \text{frequency} \times \text{wavelength}$$

Electromagnetic Radiation

There are many types of **electromagnetic radiation** including visible light. Different colours of visible light have different frequencies and wavelengths. Other electromagnetic waves have a wider range of frequencies and wavelengths outside the range of detection of human eyes; they are invisible.

All electromagnetic waves travel at the same speed through a vacuum (or air), the speed of light (300×10^6 m/s).

Longitudinal Waves

Longitudinal waves vibrate with a displacement **parallel** to the direction of travel of the wave. Examples are sound waves and seismic waves.

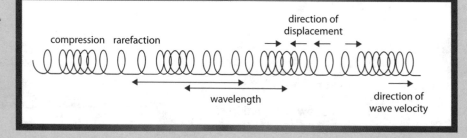

Transverse Waves

Transverse waves vibrate with a displacement **perpendicular** to the direction of travel of the wave. Examples are electromagnetic waves (including light) and water waves.

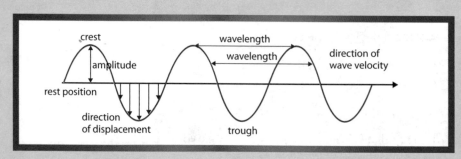

The Electromagnetic Spectrum

The electromagnetic spectrum is continuous, but waves within it are grouped into types as shown below.

Different wavelengths of electromagnetic waves are reflected, refracted, absorbed or transmitted differently by different substances and types of surfaces.

Wavelength				DECREASING WAVELENGTH				
1000 m		1 m	1 mm	0.001 mm		10^{-9} m	10^{-12} m	10^{-15} m
Long waves short waves UHF Radio and TV		microwaves	infrared			visible	ultraviolet	X-rays gamma rays

INCREASING FREQUENCY

PROGRESS CHECK

1. Define wavelength.

2. Define amplitude.

3. Which types of waves have compressions and rarefactions?

4. Which types of waves have peaks and troughs?

5. What is the speed of electromagnetic waves in a vacuum?

? EXAM QUESTION

1. Are electromagnetic waves transverse or longitudinal?

2. An electromagnetic wave has a wavelength of 0.001 mm. The speed of all electromagnetic waves is the speed of light (300×10^6 m/s). Calculate the frequency of this wave.

3. What type of wave is it likely to be?

4. Name a type of electromagnetic wave that has a shorter wavelength.

Waves in our World

Excessive exposure to **electromagnetic** waves can be detrimental. Higher frequency radiation causes greater risk.

Here are some of the dangers.

- **Infrared** waves are given off by heaters, too much can cause skin burns. Particles at the surface absorb infrared, the heat is transferred to the centre by convection or conduction. Infrared sensors are used in remote controls, burglar alarms and short distance computer data links.

- **Ultraviolet light** is emitted naturally by the Sun. Overexposure damages skin and eyes and causes cancer. High factor sunblock reduces the risks.

- **Microwaves** are absorbed by particles on the outside layers of food to a depth of about 1 cm, increasing the kinetic energy of the particles. The heat is transferred to the centre by conduction or convection. Microwaves are reflected by metal, but can go through glass and plastics. Microwaves of high frequency and energy can cause burns when absorbed by body tissue.

- **X-rays** and **gamma rays** cause mutation or destruction of cells in the body.

Seismic Waves

Seismic waves are produced by earthquakes as shock waves and can be detected by seismometers.

- P-waves (primary waves) travel through solid and liquid rock and travel faster than s-waves. They are longitudinal.

- S-waves (secondary waves) travel through solid rock only. They are transverse.

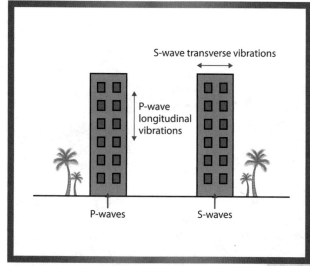

Data from seismic waves can be used to draw conclusions about the types of materials that are found inside the Earth. The outer core stops the S-waves because it is liquid. The waves are also refracted within the mantle due to the different densities of the rock.

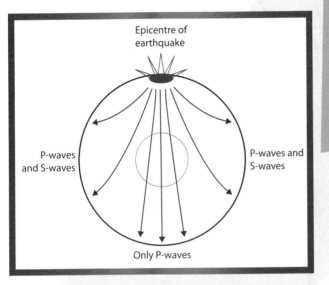

PROGRESS CHECK

1. Give **two** uses of infrared waves.

2. Explain how microwaves heat food.

3. How can we protect against ultra violet light?

4. How do gamma rays affect living cells?

5. What **two** properties are identical for all waves in a laser beam?

? EXAM QUESTION

1. Name **two** types of seismic waves produced by an earthquake.

2. Which of these is longitudinal?

3. Which of these can travel through liquid?

4. Explain how these waves can provide information about the Earth's core.

Using Waves

Mobile phones and satellites use **microwave** and **radio waves**. There is more about satellite communication on page 116.

Communication

Lasers produce a narrow intense beam of light in which all the waves are:

- the same frequency

- in phase with each other.

A laser in a CD player reflects from the surface which contains digital information in a pattern of pits.

Analogue signals have a continuously variable value.

Digital signals are either on or off.

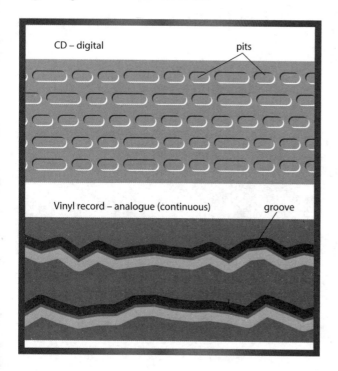

CD – digital — pits

Vinyl record – analogue (continuous) — groove

Total Internal Reflection

Visible light rays, or infrared, can be reflected along the inside of an optical fibre if the incident angle is greater than or equal to the critical angle. The critical angle is different for different materials.

For some types of glass, it is equal to about 41°, for acrylic it is equal to 42°. This effect is called **total internal reflection**.

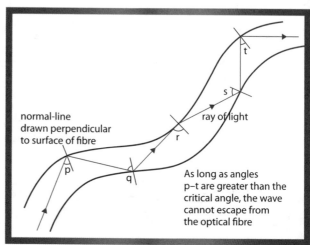

normal-line drawn perpendicular to surface of fibre

ray of light

As long as angles p–t are greater than the critical angle, the wave cannot escape from the optical fibre

- Optical fibres allow rapid transmission of data.

- The fibre can be flexible, the signal will still be internally reflected.

- The pulses of light are **digital** signals.

- Many digital pulses can be sent on the same data line (multiplexing) allowing more information to be transmitted.

- Digital signals are clearer because they create less interference.

Absorption and Reflection

Different wavelengths of electromagnetic radiation have different effects on living cells. Some pass through, some are absorbed and produce heat or an alternating current of the same frequency of the radiation.

Some cause cancerous changes and some may kill the cells. These effects depend on the type of radiation and the size of the dose.

Examples of scanning by **absorption**:

- microwaves used to monitor rain

- X-rays to see bone fractures

- ultraviolet light to detect forged bank notes.

Examples of scanning by **reflection**:

- Ultrasound to scan a foetus during pregnancy. The distance to the reflecting surface is calculated using:

$$\text{speed} = \frac{\text{distance}}{\text{time}}$$

- Optical waves for iris recognition.

Infrared uses scanning by **emission** to monitor temperature.

PROGRESS CHECK

1. Explain the difference between analogue and digital signals.

2. Some electromagnetic radiation passes through living cells, some is absorbed, some cause cancerous changes. What **two** properties of the radiation effect what happens?

3. Explain another possible effect of radiation on living cells.

4. State **one** application of scanning by absorption.

5. State **one** application of scanning by reflection.

EXAM QUESTION

The diagram shows a ray of light entering an optical fibre and reflecting off the inside surface.

a. What is this effect called?

b. Label the angle that must be greater than the critical angle for this to occur.

c. What is the value of the critical angle for glass?

d. Continue the path of the ray as it reflects along the inside of the glass fibre.

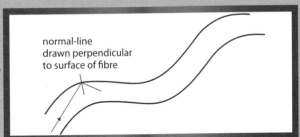

normal-line drawn perpendicular to surface of fibre

Beyond our Planet

The Universe consists of: stars and planets, comets and meteorites, black holes, large groups of stars called galaxies.

Our **galaxy** is called the Milky Way and consists of billions of stars. In our solar system, the Earth is one of many planets orbiting our star (the Sun) with different radiuses and time periods. Gravity provides the centripetal force for orbital motion.

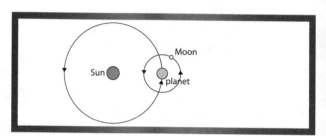

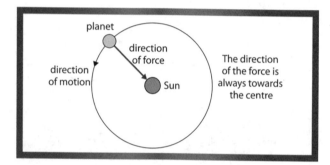

The Earth has an orbit of 365 days. Every 24 hours, it rotates on its axis creating day and night. The existence of life on our planet is determined by the position of our planet within our solar system and the position of our star, the Sun, in its life cycle.

Comets

Comets orbit the Sun with highly elliptical orbits. They are made from ice and dust from far beyond the planets. A comet has least speed when furthest from the Sun, when the gravitational pull is the weakest.

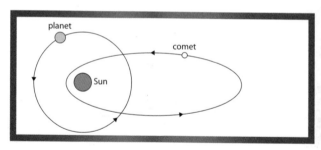

The Moon

The Moon orbits Earth. The Moon may be the remains of a planet which collided with the Earth. When two planets collide, their iron cores can merge to form a larger planet and the less dense material orbits as a Moon.

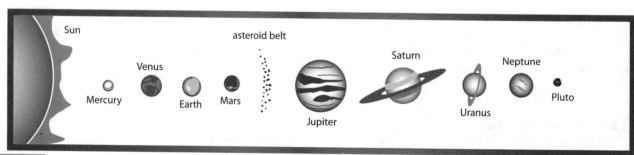

Asteroids

Asteroids are thousands of rocks orbiting the Sun, left over from the formation of the solar system. The large gravity of Jupiter disrupts the formation of a planet, causing an asteroid belt to orbit between Mars and Jupiter.

Consequences of asteroid collision can be:

- craters
- ejection of hot rocks
- widespread fires
- sunlight blocked by dust
- climate change
- species extinction (causing sudden changes of fossil numbers).

A **near-earth-object** (NEO) is an asteroid or a comet on a possible collision course with the Earth. An NEO may be seen with telescopes, monitored by satellites and deflected by explosions.

Stars

Stars are huge balls of burning gas that give off their own light. Here is the life history of a star.

- Interstellar gas cloud.
- Gravitational collapse producing a protostar.
- Thermonuclear fusion.
- A long period of normal life (**main sequence**).

At the end of its life, the outer layer of a medium weight star expands to become a **red giant**. It then drifts into space as planetary nebula leaving a **white dwarf**.

A heavy weight star also becomes a red giant, which explodes as a **supernova**, then becomes a very dense **neutron star** or a black hole. Our Sun is a medium weight main sequence star.

A **black hole** has a very large mass and very strong gravity. Nothing can escape from a black hole, not even light. Much of the universe is made up of **dark matter**, we know very little about its nature.

PROGRESS CHECK

1. How often does the Earth rotate?

2. What is the name of our galaxy?

3. What is a red giant?

4. Describe the end of a life cycle of a massive star.

5. Why can light not escape from a black hole?

EXAM QUESTION

Our solar system consists of nine planets orbiting the Sun.

a. Name **two** other things that orbit the Sun.

b. Why is the asteroid belt between Mars and Jupiter?

c. Give **three** possible consequences of collision with an asteroid.

d. What is a Moon?

Communication

Mobile phones and satellites use **microwave** and **radio waves**. Signals can be reflected from the ionosphere or received and re-transmitted by satellites.

Communication Signals

Diffraction, refraction and interference around obstacles cause signal loss.

This can be reduced by:

- limiting the distance between transmitters
- positioning transmitters as high as possible.

Transmission dishes also experience signal loss due to diffraction.

There have been claims that using mobile phones or living near a phone mast may be dangerous.

Magnetic Fields

The Earth is surrounded by a magnetic field. This is because the Earth's core contains a lot of molten iron. Magnets have a North and a South pole which can be identified using a plotting compass.

The magnetic field around our planet shields us from much of the ionising radiation from space. An electrical current in a coil also creates a magnetic field.

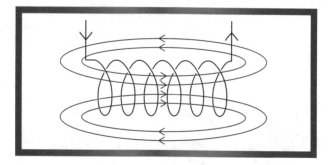

A current flows in the coil (also called a solenoid) producing a magnetic field. The field is the same shape as the field around a bar magnet. As long as the current flows, the coil acts like a bar magnet.

Cosmic Rays

Cosmic rays causing ionising radiation and **solar flares** from the Sun also interfere with the operation of **artificial satellites**.

Cosmic rays:

- are fast moving particles which create gamma rays when they hit the atmosphere
- spiral around the Earth's magnetic field to the poles
- cause the Aurora Borealis.

Solar Flares

Solar flares:

- are clouds of charged particles from the Sun
- are ejected at high speed
- produce strong disturbed magnetic fields.

Artificial Satellites

Artificial satellites are used for:

- telecommunications
- weather prediction
- spying
- satellite navigation systems.

PROGRESS CHECK

1. What types of waves do mobile phones use?
2. Give **two** ways that signal loss can be reduced.
3. Can living near a phone mast be dangerous?
4. State **two** sources of artificial satellite interference.
5. Suggest **two** uses for artificial satellites.

? EXAM QUESTION

The following diagram shows a coil of wire that has a current passing through it.

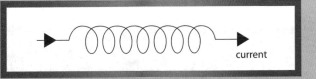

current

a. Sketch the shape of the magnetic field around the coil of wire.

b. What does the magnetic field around the Earth protect us from?

c. How does the Earth's magnetic field affect cosmic rays?

d. Name **one** other source of magnetic fields in space.

Space Exploration

Space has no gravity, no atmosphere and varies dramatically in temperature.

Extended periods in space can cause deterioration of bones and heart and can subject an astronaut to dangerous levels of radiation.

A manned spacecraft must provide:

- enough fuel
- shielding from cosmic rays
- heating and cooling
- artificial gravity
- air supply
- enough food and water
- exercise machines.

Space travel takes a long time and covers huge distances. **Light years** are used to measure these distances, one light year is the distance light travels in a year.

An unmanned spacecraft can withstand conditions that humans cannot. They have the advantages of cost and safety. However, if something goes wrong, it is difficult to fix. Unmanned spacecraft can collect information on:

- temperature
- magnetic field
- radiation
- gravity
- atmosphere.

Mass, Weight and Force

The **mass** of a body is how much matter it is made of, it does not change. Mass is measured in kilograms (kg). The **weight** of a body is the amount of gravitational **force** acting on it and can be calculated as follows:

$$\text{Weight (N)} = \text{mass (kg)} \times \text{acceleration of free fall (N/kg)}$$

The acceleration of free fall (g) is about 9.8 N/kg on Earth.

A spacecraft is powered by ejecting fuel backwards. The **action** of the fuel backwards causes the **reaction** of the spacecraft forwards. Its motion can be predicted using the equation:

$$\text{Force (N)} = \text{mass (kg)} \times \text{acceleration (m/s/s)}$$

Observations of our universe can be carried out on Earth or in space. Telescopes can detect visible light or other non-visible electromagnetic waves.

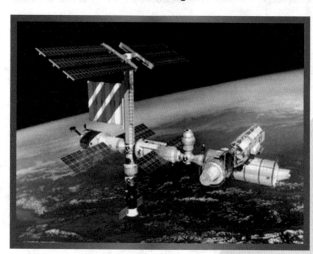

The Big Bang Theory

One of the theories of the origin of the universe is that it began with a **Big Bang** from a very small initial point.

When galaxies move away from us at high speeds, light waves from them become stretched out, moving their wavelength closer to the red end of the visible spectrum, called 'red-shift'. The more distant galaxies have a greater red-shift.

This means that the galaxies are moving away from each other and the universe is expanding. This supports the Big Bang theory and predicts that the universe began about 15 000 million years ago.

The faint microwave radiation from space, known as cosmic rays and picked up by telescopes, may be left over from the Big Bang.

It is thought that the universe may eventually stop expanding and reach a **steady state**, or that it might eventually collapse and then expand again, **oscillating** continuously.

👁 PROGRESS CHECK

1. How can space travel affect the human body?

2. State **three** things a manned spacecraft must provide.

3. What is a light year?

4. Give **three** things that an unmanned spacecraft can collect information about.

5. What **two** pieces of evidence are there to support the Big Bang theory?

❓ EXAM QUESTION

1. Explain the difference between mass and weight.

2. What is the weight on Earth of an object of mass 7.5 kg? (g = 9.8 N/kg on Earth)

3. The object is taken into space in a rocket: state the effects, if any, on the mass and the weight of the object.

4. The rocket is propelled forwards by the reaction force caused by the rocket's fuel. Which direction is the rocket's fuel projected?

Answers

Day 1

pages 4–5
How Science Works
PROGRESS CHECK

1. The variable we choose to change in an experiment
2. The variable that we measure in an experiment
3. A variable that can be put in order
4. A variable that can have any whole number value
5. It is close to the true value

EXAM QUESTION

a. The force applied
b. The length of the spring
c. Using the same spring etc.

Biology
pages 6–7
Healthy Living
PROGRESS CHECK

1. Fats, carbohydrates, proteins, vitamins, minerals, fibre and water
2. Glucose + oxygen → carbon dioxide + water + **ENERGY**
3. Lactic acid

EXAM QUESTION

1. Saturated fats
2. Statin drugs and eating polyunsaturated fats
3. Genetic factors, smoking and alcohol

pages 8–9
Central Nervous System
PROGRESS CHECK

1. Central Nervous System
2. Sensory, motor and relay neurone
3. Voluntary – you have to think about, under conscious control. Reflex – rapid, involuntary response

EXAM QUESTION

1. Receptor, relay, effector
2. By transmission of a chemical across a synapse
3. Stimulus

pages 10–11
The Eye
PROGRESS CHECK

1. Binocular – eyes at the front, monocular – eyes at the side
2. Ciliary muscles
3. Fat and round

EXAM QUESTION

1. Iris – controls how much light enters the eye, lens – changes shape to focus, suspensory ligaments – holds lens in place
2. Light passes through the cornea and lens and is focused on the retina
3. Stiff suspensory ligaments, weak ciliary muscles, causes problems when judging distances

pages 12–13
The Brain
PROGRESS CHECK

1. Medulla
2. Epilepsy
3. Cerebellum

EXAM QUESTION

1. Neurones die causing paralysis and loss of speech
2. Epilepsy, Parkinson's disease and brain tumours (any two)
3. Learning increases the likelihood of nerve pathways transmitting impulses

pages 14–15
Homeostasis
PROGRESS CHECK

1. Maintaining a constant internal environment
2. (i) glucagons (ii) insulin
3. The pancreas not making enough insulin

EXAM QUESTION

a. Vasoconstriction of blood vessels, shivering, stop producing sweat, increased respiration
b. Hypothalamus
c. 37°C

pages 16–17
Controlling Fertility
PROGRESS CHECK

1. Ovaries
2. FSH, LH
3. Release of an egg from the ovary

EXAM QUESTION

1. FSH, oestrogen, LH, progesterone
2. Oestrogen
3. FSH

pages 18–19
Pathogens
PROGRESS CHECK

1. Bacteria, virus, fungi, protozoa
2. Pathogens
3. Transfers pathogens from one organism to another, e.g. a mosquito

EXAM QUESTION

1. Bacterial
2. Reproduce inside living cells, release lots of other viruses, killing the cell in the process
3. Virus, fungi, protozoa (any two)

pages 20–21
Natural Defence
PROGRESS CHECK

1. Antitoxins
2. They have antigens on their surface
3. Sterilising equipment, disinfectants, antiseptics, general good hygiene

EXAM QUESTION

1. Engulf bacteria and destroy them
2. Lymphocytes recognise antigens and produce antibodies that attach to the antigens and destroy them/help phagocytes destroy them
3. Antibodies are specific to particular antigens

Day 2
pages 22–23
Artificial Immunity
PROGRESS CHECK

1. Penicillin
2. They mutate
3. Where neither the patient nor the doctor know who is receiving the drug

EXAM QUESTION

1. Antibiotics only fight bacteria
2. Antibodies generated from vaccines are specific and will not fight a cold as it is a different virus from 'flu
3. Passive

pages 24–25
Drugs
PROGRESS CHECK

1. Lungs, brain, kidney, liver
2. Nicotine, tar, carbon monoxide
3. Emphysema, bronchitis, heart and blood vessel problems, lung cancer

EXAM QUESTION

a. Solvents – slow down breathing and heart rates, damage kidney and liver
b. Alcohol – cirrhosis of the liver
c. Stimulants – speed up the brain and nervous system
d. Sedatives – slow down the brain

pages 26–27
Genes and Chromosomes
PROGRESS CHECK

1. 46
2. If it affects reproductive cells
3. Sexual reproduction, the environment

EXAM QUESTION

DNA – chromosomes are made up of this
Alleles – alternative form of a gene
Genes – instructions for characteristics
Chromosomes – there are 23 of these in human gametes

pages 28–29
Genetic Engineering
PROGRESS CHECK

1. Identifying all the genes in human DNA and studying them
2. Where genes from one organism are removed and inserted into cells of another

3. Inserting a non-disease-causing gene into the body cells of a person with the disease

EXAM QUESTION

1. Pancreas
2. The insulin-making gene is cut out of the pancreas using enzymes and inserted into a plasmid of a bacterium. The plasmid is put back into the bacterium where it multiplies and produces insulin in a fermenter
3. They could mutate into harmful bacteria

pages 30–31
Inheritance and Disease
PROGRESS CHECK

1. By inheriting two recessive alleles
2. A disorder of the red blood cells that affects the oxygen carrying capacity of the blood
3. Carriers are resistant to malaria

EXAM QUESTION

a.

	R	r
R	RR	Rr
r	Rr	rr

b. (i) RR or Rr (ii) rr

pages 32–33
Selective Breeding
PROGRESS CHECK

1. Asexual reproduction
2. Clones of each other but not their parents
3. Lots of calves can be produced from one bull and one cow in the time it normally takes to produce one. The farmer does not have to own the cow or the bull

EXAM QUESTION

a. False, if it involves sexual reproduction
b. Cuttings or tissue culture
c. It always involves sexual reproduction which produces variation

pages 34–35
The Environment
PROGRESS CHECK

1. Rear animals (or plants) intensively, reduce stages in the food chain
2. Pyramid of numbers
3. Pyramid of biomass

EXAM QUESTION

1. Carbon dioxide, glucose
2. By measuring how much oxygen is produced in a given time
3. Amount of carbon dioxide, amount of light and temperature

Day 3
pages 36–37
Environmental Damage
PROGRESS CHECK

1. Burning fossil fuels
2. Sulfur dioxide and nitrogen dioxides
3. Carbon dioxide and methane

EXAM QUESTION

1. Chopping down trees
2. Soil erosion, destruction of habitats, less rainfall, less food, disruption to food chains
3. Trees absorb carbon dioxide from the air and burning the timber also increases carbon dioxide in the air

pages 38–39
Ecology and Classification
PROGRESS CHECK

1. Carl Linnaeus
2. 5
3. Animals with a backbone

EXAM QUESTION

1a. A group of living things able to breed together to produce fertile offspring
1b. Leo
2. Randomly placed quadrat, plants are counted if more than half are in the quadrat or touching the frame, process repeated and then an average number of species in a field is calculated

pages 40–41
Adaptation and Competition
PROGRESS CHECK

1. Thick coat for camouflage and warmth
2. Where an organism lives that provides it with the conditions necessary for survival
3. The predator/prey cycle keeps numbers constant

EXAM QUESTION

1. Hot and dry
2. To reduce water losses – round shape, thick cuticles, and spines. It can store water and has long roots to reach water

pages 42–43
Evolution

PROGRESS CHECK

1. Natural selection
2. Fossils
3. Only the best adapted to a change in the environment will survive and breed

EXAM QUESTION

a. Natural selection bring about a mutation
b. It was camouflaged in polluted city areas
c. Adapt

Chemistry

pages 44–45
Air and Air Pollution

PROGRESS CHECK

1. Nitrogen
2. Carbon dioxide
3. Plants evolved/photosynthesis
4. Oxygen
5. Sulfur dioxide

EXAM QUESTION

a. Mars/Venus
b. Plants evolved/photosynthesis
c. Global warming/greenhouse effect

pages 46–47
Crude Oil

PROGRESS CHECK

1. A compound that only contains carbon and hydrogen atoms
2. A part of crude oil/ a group of molecules with a similar number of carbon atoms
3. At the top
4. Bitumen
5. Kerosene

EXAM QUESTION

a. Fractional distillation
b. Evaporates
c. Condense
d. Fractions

pages 48–49
Food Additives

PROGRESS CHECK

1. Throughout the European Union
2. Keeps unblendable liquids mixed together
3. Natural
4. Colouring
5. Antioxidant

EXAM QUESTION

1. Sodium hydrogen carbonate
2. To improve the texture
3. Carbon dioxide

pages 50–51
Vegetable Oils

PROGRESS CHECK

1. A and D
2. Sunflower oil and olive oil etc.
3. Seeds, nuts and fruits
4. They have carbon double bonds
5. Vinegar and vegetable oil

EXAM QUESTION

a. Emulsion
b. Polyunsaturated
c. Biofuel
d. Bromine water

pages 52–53
Fuels

PROGRESS CHECK

1. Hydrogen and carbon
2. Carbon dioxide and water vapour
3. Carbon
4. Global warming/greenhouse effect
5. It produces soot/ carbon monoxide

EXAM QUESTION

1. Carbon dioxide
2. Carbon/soot/carbon monoxide/ carbon dioxide
3. Water (vapour)

Day 4
pages 54–55
Alkanes and Alkenes

PROGRESS CHECK

1. C_nH_{2n+2}
2. Fuels
3. Covalent bonds
4. Cracking
5. C_2H_4

EXAM QUESTION

a. Ethene
b. Alkenes
c. C_nH_{2n}

pages 56–57
Limestone

PROGRESS CHECK

1. Calcium carbonate
2. Sedimentary
3. Calcium oxide (quicklime) and carbon dioxide
4. Quicklime
5. Slaked lime

EXAM QUESTION

1. Igneous
2. Marble

pages 58–59
Cosmetics

PROGRESS CHECK

1. A mixture made when a solvent dissolves a solute
2. Nail varnish etc.
3. Ethanol
4. Jasmine/lavender etc.
5. Reacting an alcohol with a carboxylic acid

EXAM QUESTION

a. Fermentation
b. ethene + steam → ethanol

pages 60–61
Structure of the Earth

PROGRESS CHECK

1. Solid
2. Liquid
3. Mantle
4. Dozen/twelve
5. Tsunami

EXAM QUESTION

a. The crust and upper mantle
b. A few centimetres per year
c. Mountains formed as the Earth's crust shrank as it cooled down

pages 62–63
New Materials

PROGRESS CHECK

1. It is too big
2. Nylon
3. Teflon
4. To make non-stick saucepans
5. Bullet-proof vests

EXAM QUESTION

a. It contains very small fibres
b. The small fibres strap air. This acts as a layer of insulation which stops body heat from being lost

pages 64–65

Pollution

PROGRESS CHECK

1. New jobs
2. Noise/dust/lorries
3. Plastic
4. It can produce toxic gases
5. It is hard to separate them out

EXAM QUESTION

a. It can absorb water
b. It is difficult for bacteria/mould to grow so food stays fresh for longer

pages 66–67

Useful Metals

PROGRESS CHECK

1. It is too soft
2. Alloy
3. Copper and zinc
4. Drinks cans/bicycles/aeroplanes etc.
5. Rutile

EXAM QUESTION

a. It forms a layer of aluminium oxide which prevents any further reaction
b. The car body will be lighter so the car will have a better fuel economy/the car body will corrode less so it may last for longer
c. An aluminium car body will be more expensive to produce

Day 5

pages 68–69

Iron and Steel

PROGRESS CHECK

1. Iron (III) oxide
2. Gold
3. Blast furnace
4. Reduction
5. 96%

EXAM QUESTION

a. Wrought iron
b. Low carbon steel
c. Cast iron
d. Iron (III) oxide

pages 70–71

Salts

PROGRESS CHECK

1. The production of fertilisers/as colouring agents/in fireworks
2. Barium sulfate
3. Barium chloride +sodium sulfate → barium sulfate + sodium chloride
4. Solid
5. Aqueous

EXAM QUESTION

a. It is used to grit roads
b. Hydrogen
c. Sterilise water etc.

pages 72–73

The Periodic Table

PROGRESS CHECK

1. Halogens/group 7
2. Noble gases/group 0
3. Alkali metals/ group 1
4. Balloons
5. Filament lamps

EXAM QUESTION

a. Halogens/group 7
b. Chlorine + potassium bromide → potassium chloride + bromine

pages 74–75

Atomic Structure

PROGRESS CHECK

1. They have the same number of protons
2. Nucleus
3. Electrons
4. In a periodic table
5. Roughly 100

EXAM QUESTION

a. calcium carbonate + hydrochloric acid → calcium chloride + water + carbon dioxide
b. One carbon atom and two oxygen atoms

pages 76–77

Chemicals

PROGRESS CHECK

1. Hydrogen
2. Oxidising chemical
3. Toxic
4. Attack and destroy living tissue
5. Orange/yellow

EXAM QUESTION

a. Carbon dioxide
b. Oxygen
c. Chlorine
d. Ammonia

pages 78–79

Rates of Reaction

PROGRESS CHECK

1. They need to collide and when they do collide they must have enough energy to react (activation energy)
2. Increases the rate of reaction
3. Decreases the rate of reaction
4. Increases the rate of reaction
5. At the start

EXAM QUESTION

a. Catalysts
b. Increases
c. Decreases
d. Reactants

pages 80–81

Energy

PROGRESS CHECK

1. Exothermic reaction
2. Endothermic
3. Energy
4. Calorimetry
5. J

EXAM QUESTION

a. Exothermic
b. 84 J

Day 6
Physics
pages 82–83

Producing Current

PROGRESS CHECK

1. How much charge flows in a certain amount of time
2 a. True
   b. False
   c. False – ions do not move
   d. True

EXAM QUESTION

a. More turns, greater speed, stronger magnet
b. AC
c. DC is always in one direction, AC changes direction
d. Ammeter in series with (next to) lamp

pages 84–85

Resistance

PROGRESS CHECK

1. $1.5\,\Omega$
2. The current will decrease
3.

4. As the current increases the filament lamp gets hot and the resistance increases
5. Straight line

EXAM QUESTION

a. $6.0\,\Omega$
b. The resistance increases
c. The current will decrease
d. In a fire alarm

pages 86–87

Power

PROGRESS CHECK

1. $1.5\,W$
2. $240\,000\,J$
3. $0.067\,kWhs$
4. $4.8\,A$
5. 96p

EXAM QUESTION

a. $144\,000\,J$, $0.04\,KWhs$
b. 28.8p
c. $0.005\,KWhs$
d. 3.6p

pages 88–89

Using Electricity

PROGRESS CHECK

1. They have decreased in size, allowing computers to operate at higher speeds
2. Sending many signals along the same cable at once
3. They have very little friction
4. The resistance of some materials reduces to almost zero at very low temperatures
5. A short piece of thin wire that melts and breaks the circuit if the current is too high

EXAM QUESTION

a. Electricity to kinetic energy and heat
b. A residual current circuit breaker, it disconnects the circuit when the flow of current is too high
c. If a loose wire touches the outside casing the earth wire allows a large current to flow to earth, blowing the fuse and breaking the circuit
d. The casing is plastic, insulating the user from any loose wires

pages 90–91

Harnessing Electricity

PROGRESS CHECK

1. 19V
2. Increasing the voltage decreases the current and loses less energy to heat
3. Up to $400\,000\,V$
4. They do not spoil the landscape
5. Any three of:
   a. Increasing the number of turns on the coil
   b. Increasing the current
   c. Using a stronger magnet
   d. Using a bigger coil

EXAM QUESTION

a. 10:1
b. and d

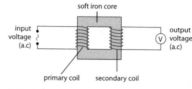

c. a step down

pages 92–93

Harnessing Energy 1

PROGRESS CHECK

1. Chemical energy to heat to kinetic energy (in a turbine which turns a generator) to electrical energy
2. Solar, wind power and moving water
3. No pollution
4. It can disturb habitats
5. Heat to kinetic to electrical

EXAM QUESTION

a. Renewable
b. It requires no fuel, it causes no pollution
c. Depends on wind speed, can disturb habitats
d. Kinetic energy of the wind to electrical energy

pages 94–95

Harnessing Energy 2

PROGRESS CHECK

1. Require large areas of land
2. Sulfur dioxide
3. Carbon dioxide
4. CFCs
5. It reflects radiation from the city, causing warming

EXAM QUESTION

a. Uranium
b. It does not contribute to global warming, small amounts of fuel give large amounts of energy
c. High maintenance costs, waste material disposal is expensive
d. The waste can cause cancer if not disposed of correctly, risk of accidental emission of radioactive material

pages 96–97

Keeping Warm

PROGRESS CHECK

1. Convection, conduction and radiation
2. Conduction
3. It expands and becomes less dense
4. Matt black
5. Loft insulation, double glazing, curtains

EXAM QUESTION

a. Conduction
b.

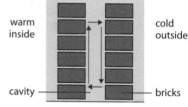

c. By filling the cavity with insulation
d. Cool

pages 98–99

Efficiency

PROGRESS CHECK

1. $\text{Efficiency} = \dfrac{\text{useful energy output}}{\text{total energy input}} \times 100\%$

2. 50%
3. 15%
4. They lose 75% of their energy as heat
5. 35%

EXAM QUESTION

a. Power is energy per second
b. Electricity
c. 8.3%
d. It is lost as heat to the surroundings

Day 7
pages 100–101

Heat

PROGRESS CHECK

1. To absorb more energy
2. A measurement of hotness and is measured in degrees Celsius (°C)
3. Heat is a measurement of energy, measured in Joules
4. They have a large surface area compared with their mass so lose heat quickly
5. Tiles are good thermal conductors and transfer heat away quickly

EXAM QUESTION

a. They are knocked loose from the silicone crystals
b. Surface area and light intensity
c. They can operate in remote locations
d. They are expensive

pages 102–103

Heat Calculations

PROGRESS CHECK

1. How much energy 1kg of a material needs to increase (or loses when decreasing) its temperature by 1 °C. The unit of specific heat capacity is Joules per kg per degree
2. At normal pressure water freezes and melts at 0 °C
3. The specific latent heat of a substance is the amount of energy required to change the state of 1 kg of that substance at constant temperature
4. 42 000 J
5. 1925 J

EXAM QUESTION

a. The water needs energy to change state at constant temperature
b. 420 000 J
c. Some heat will be lost to the surroundings
d. 2 260 000 J

pages 104–105

Radiation

PROGRESS CHECK

1. Neutral – no charge
2. If a nucleus is unstable a neutron may spontaneously change into a proton inside the nucleus. When it does this, it emits an electron
3. Huge distances
4. Deflected a lot due to their low mass
5. The ability of radiation to cause other particles to gain or lose electrons

EXAM QUESTION

a. Isotopes
b. A helium nucleus, two protons and two neutrons
c. Alpha particles only travel a few cm in air
d. Place the alpha source a few cm from the detector. Either move the detector to a distance of about 1.0 m or place a thick sheet of paper between the source and the detector. Compare the readings

pages 106–107

Radiation in Our World

PROGRESS CHECK

1. Rocks, cosmic rays and human activity
2. The time it takes for the number of nuclei in a sample to halve
3. It is highly ionising and very dangerous if it is taken into the body in food or by breathing in radioactive gas or dust. Once absorbed, the radiation can cause damage deep inside the body
4. Smoke detectors
5. Sterilising equipment

EXAM QUESTION

a.

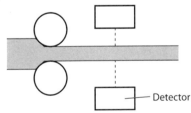

Detector

b. Beta
c. As a tracer to detect underground leaks in pipes
d. Use tongs or gloves, minimise the exposure time

pages 108–109

Waves

PROGRESS CHECK

1. The distance between any point on a wave and the same point on the next wave
2. The maximum displacement of the waves from rest position
3. Longitudinal
4. Transverse
5. The speed of light (300×10^6 m/s)

EXAM QUESTION

1. Transverse
2. 300×10^{14} Hz
3. Infrared
4. X-rays

pages 110–111

Waves in our World

PROGRESS CHECK

1. Remote controls, burglar alarms
2. They are absorbed by particles on the outside layers of food to a depth of about 1 cm. The heat is transferred to the centre by conduction or convection
3. Wear high factor sunblock
4. They cause mutation or destruction
5. Frequency and phase

EXAM QUESTION

1. P-waves and S-waves
2. P-waves
3. P-waves
4. Only P-waves can travel through liquid, so when S-waves are stopped by parts of the core, we know that these parts must be liquid

pages 112–113

Using Waves

PROGRESS CHECK

1. Analogue signals have a continuously variable value, digital signals are either on or off
2. The type of radiation and the size of the dose
3. Can power an alternating current of the same frequency as the radiation
4. Ultraviolet light to detect forged bank notes
5. Optical waves for iris recognition

EXAM QUESTION

a. Total internal reflection
b.

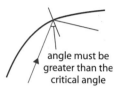

angle must be
greater than the
critical angle

c. About 41°
d.

pages 114–115

Beyond our Planet

PROGRESS CHECK

1. Every 24 hours
2. The Milky Way
3. A star that has expanded outer layers near the end of its life
4. It becomes a red giant, then explodes as a supernova, then becomes a very dense neutron star or a black hole
5. Because of the very strong gravity

EXAM QUESTION

a. Comets and asteroids
b. Because the large gravity of Jupiter disrupts the formation of a planet
c. Craters, sunlight blocked by dust, climate change
d. An object that orbits a planet

pages 116–117

Communication

PROGRESS CHECK

1. Microwaves and radio waves
2. Limiting the distance between transmitters, positioning transmitters as high as possible
3. There have been claims that it is
4. Cosmic rays causing ionising radiation and solar flares from the Sun
5. Spying, satellite navigation systems

EXAM QUESTION

a.

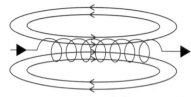

b. Ionising radiation from space
c. They spiral around the magnetic field to the poles, causing Aurora Borealis
d. Solar flares

pages 118–119

Space Exploration

PROGRESS CHECK

1. It can cause deterioration of bones and heart
2. Artificial gravity, air supply, enough food and water
3. The distance light travels in a year
4. Temperature, magnetic field, atmosphere
5. The expanding universe and cosmic rays

EXAM QUESTION

1. Mass is the amount of matter something is made of and is measured in kilograms. The weight of a body is a measure of the amount of gravitational force acting on it and is measured in Newtons
2. 73.5 N
3. The mass does not change, the weight decreases
4. Backwards

Notes

Notes